T0292382

Methods for Measuring Greenhouse Gas Balances and Evaluating Mitigation Options in Smallholder Agriculture

Todd S. Rosenstock • Mariana C. Rufino
Klaus Butterbach-Bahl • Eva Wollenberg
Meryl Richards
Editors

Methods for Measuring Greenhouse Gas Balances and Evaluating Mitigation Options in Smallholder Agriculture

Editors
Todd S. Rosenstock
World Agroforestry Centre (ICRAF)
Nairobi, Kenya

Klaus Butterbach-Bahl
Karlsruhe Institute of Technology, Institute of Meteorology and Climate Research
Atmospheric Environmental Research (IMK-IFU)
International Livestock Research Institute (ILRI)
Nairobi, Kenya

Meryl Richards
University of Vermont
CGIAR Research Program on Climate Change, Agriculture and Food Security (CCAFS)
Burlington, VT, USA

Gund Institute for Ecological Economics
University of Vermont
Burlington, VT, USA

Mariana C. Rufino
Center for International Forestry Research (CIFOR)
Nairobi, Kenya

Eva Wollenberg
University of Vermont
CGIAR Research Program on Climate Change, Agriculture and Food Security (CCAFS)
Burlington, VT, USA

Gund Institute for Ecological Economics
University of Vermont
Burlington, VT, USA

ISBN 978-3-319-29792-7 ISBN 978-3-319-29794-1 (eBook)
DOI 10.1007/978-3-319-29794-1

Library of Congress Control Number: 2016933777

Printed on acid-free paper

This Springer imprint is published by SpringerNature
The registered company is Springer International Publishing AG Switzerland

Foreword

In this book, the author team describe concepts and methods for measurement of greenhouse gas emissions and assessment of mitigation options in smallholder agricultural systems, developed as part of the SAMPLES project. The SAMPLES (Standard Assessment of Agricultural Mitigation Potential and Livelihoods) system adapts existing internationally accepted methodologies to allow a range of stakeholders to assess greenhouse gas (GHG) emissions from different agricultural activities, to identify how these emissions might be reduced (i.e., mitigation), and to provide data through an online dataset that can be used to aid in these efforts.

The book is divided into three sections: (1) designing a measurement program to allow users to identify what measurements are needed and how to go about taking the measurements, (2) data acquisition, describing how to deal with complex issues such as land use change, and (3) identifying mitigation options, which deals with scaling issues, how to use models, and how to assess trade-offs. Within each section is a series of chapters, written by leading experts in the field, providing clear guidelines on how to deal with each of the issues raised.

The work was begun at an international workshop in 2012, and the authors have since produced this synthesis. Through this work, the authors provide a comprehensive and transparent system to allow stakeholders to calculate and reduce agricultural GHG emissions, and assess other impacts. Since it builds on established and internationally accepted methodologies it is robust, yet the authors have managed to break down the complex and potentially overwhelming concepts and methods into bite-sized chunks. Difficult subjects such as inaccuracy and uncertainty are not avoided, yet the authors manage to make these topics accessible and the process manageable.

Potential users include, but are not limited to, national agricultural research centers, developers of national and subnational mitigation plans that include agriculture, agricultural commodity companies and agricultural development projects, and students and instructors. Anyone with an interest in agriculture, greenhouse gas emissions, and how to minimize these emissions will find the book immensely useful.

Pete Smith

Preface

In October 2011, we faced a problem. We knew that the greenhouse gas (GHG) emissions from smallholder agriculture contributed to climate change and could present a climate change mitigation solution; however, we had no idea by how much. Experts at a workshop on farm and landscape GHG accounting organized by the CGIAR Research Program on Climate Change, Agriculture and Food Security (CCAFS) and the Food and Agriculture Organization of the UN (FAO) quickly realized that there were few data to support GHG quantification in smallholder systems. Compounding the issue, everyone seemed to use different approaches for estimating emissions and mitigation impacts. This meant that even if data were available they could not easily be compared. We needed to harmonize methods. However, the available measurement protocols typically focused on singular farming activities, such as soil fluxes or biomass. This contrasted with the realities of diverse smallholder farms, which have multiple greenhouse gas sources and sinks. We needed a more holistic approach that could capture the diversity and complexity of smallholder systems.

To meet these challenges, workshop participants conceived the idea for the SAMPLES (Standard Assessment of Agricultural Mitigation Potential and Livelihoods) project, which CCAFS initiated in 2012, in collaboration with partners at FAO's Mitigation of Climate Change in Agriculture (MICCA) program, the Global Research Alliance for Agricultural Greenhouse Gas Emissions (GRA), and multiple universities worldwide. The goal of SAMPLES was to increase and improve the availability of data on greenhouse gas emissions and removals in smallholder agricultural systems and to design ways to reduce the cost and improve the quality of future data collection efforts for these systems, especially to quantify the impacts of low emissions practices. SAMPLES has worked toward these objectives through four interrelated activities: (1) global emission hotspot analysis, (2) estimating emissions and potential reductions in a whole-farm context, (3) capacity building around GHG quantification, and (4) policy engagement.

This volume is the product of 3 years of work toward creating a coherent approach and dataset on smallholder farm emissions and mitigation options. The SAMPLES quantification framework was developed during an expert workshop on

GHG quantification held in Garmisch-Partenkirchen, Germany, in October 2012 and hosted by the Karlsrühe Institute of Technology. Following the workshop, authors reviewed the available "best practice" in greenhouse gas quantification methods and in some cases developed new methods to adapt the approach to the research constraints found in developing countries. Methods described herein are based on internationally accepted methods and have been reviewed by experts in the field.

These guidelines are intended to inform the field measurements of agricultural GHG sources and sinks, especially to assess low emissions development options in smallholder agriculture in tropical developing countries. The methods provide a standard for consistent, robust data that can be collected at reasonable cost with available equipment. They can be used to support improved emissions factors for country inventories, to assess the mitigation impacts of projects, or as methods for scientific studies. The accompanying website (http://samples.ccafs.cgiar.org/) provides additional resources such as links to step-by-step guidelines, scientific publications, and a database of agricultural emission factors.

We acknowledge with gratitude the following individuals who helped conceive this volume at a workshop in Garmisch-Partenkirchen, Germany, in October 2012:

Alain Albrecht, Institut de Recherche pour le Développement (IRD), France
Andre Butler, IFMR LEAD, India
Klaus Butterbach-Bahl, International Livestock Research Institute (ILRI) and Institute of Meteorology and Climate Research Atmospheric Environmental Research (IMK-IFU)
Aracely Castro Zuñiga, Independent Consultant, Italy
Ngonidzashe Chirinda, International Center for Tropical Agriculture (CIAT), Colombia
Alex DePinto, International Food Policy Research Institute (IFPRI), USA
Jonathan Hickman, Columbia University, USA
ML Jat, International Maize and Wheat Improvement Center (CIMMYT), India
Brian McConkey, Agriculture and Agri-food Canada and Global Research Alliance on Agricultural Greenhouse Gas Emissions, Canada
Ivan Ortiz Monasterio, International Maize and Wheat Improvement Center (CIMMYT), Mexico
Barbara Nave, BASF, Germany
An Notenbaert, International Livestock Research Institute (ILRI), Kenya
Susan Owen, Center for Ecology and Hydrology, UK
JVNS Prasad, Central Research Institute for Dryland Agriculture (CRIDA), India
Meryl Richards, University of Vermont and CGIAR Research Program on Climate Change, Agriculture and Food Security (CCAFS), USA
Philippe Rochette, Agriculture and Agri-Food Canada
Todd Rosenstock, World Agroforestry Centre (ICRAF), Kenya
Mariana Rufino, Center for International Forestry Research (CIFOR), Kenya

Björn Ole Sander, International Rice Research Institute (IRRI), Philippines
Sean Smukler, University of British Columbia, Canada
Piet van Asten, International Institute of Tropical Agriculture (IITA), Uganda
Mark van Wijk, International Livestock Research Institute (ILRI), Costa Rica
Jonathan Vayssieres, CIRAD, Senegal
Eva Wollenberg, University of Vermont and CGIAR Research Program on Climate Change, Agriculture and Food Security (CCAFS), USA
Xunhua Zheng, Institute of Atmospheric Physics-Chinese Academy of Sciences (IAP-CAS), China

We also acknowledge the following individuals and organizations that provided feedback on all or part of the guidelines during the review process:

Juergen Augustin, Leibniz Centre for Agricultural Landscape Research, Germany
Rolando Barahona Rosales, National University of Colombia (Medellín), Colombia
Ed Charmley, Commonwealth Scientific and Industrial Research Organisation, Australia
Nicholas Coops, University of British Columbia, Canada
Nestor Ignacio Gasparri, National University of Tucumán, Argentina
Jeroen Groot, Wageningen University and Research Centre, Netherlands
Ralf Kiese, Karlsruhe Institute for Technology, Germany
Brian McConkey, Agriculture and Agri-Food Canada
Eleanor Milne, Colorado State University, USA
Carlos Ortiz Oñate, Technical University of Madrid, Spain
David Powlson, Rothamsted Research, UK
Philippe Rochette, Agriculture and Agri-Food Canada
Don Ross, University of Vermont, USA
Sileshi Weldesmayat, World Agroforestry Centre, Kenya
Jonathan Wynn, University of South Florida, USA
Christina Seeberg-Elverfeldt, German Federal Ministry of Economic Cooperation and Development (BMZ), Germany
Marja-Liisa Tapio-Biström, Ministry of Agriculture and Forestry, Finland
Kaisa Karttunen, Agriculture and Development Consultant, Finland
The Mitigation of Climate Change in Agriculture (MICCA) Program of the United Nations Food and Agriculture Organization.

This work was undertaken as part of the CGIAR Research Program on Climate Change, Agriculture and Food Security (CCAFS), which is a strategic partnership of CGIAR and Future Earth. This research was carried out with funding by the European Union (EU) and with technical support from the International Fund for Agricultural Development (IFAD). The views expressed in the document cannot be taken to reflect the official opinions of CGIAR, Future Earth, or donors.

The CGIAR Research Program on Climate Change, Agriculture and Food Security (CCAFS) is supported by Australia (ACIAR), the Government of Canada through the Federal Department of the Environment, Denmark (DANIDA), Ireland

(Irish Aid), the Netherlands (Ministry of Foreign Affairs), New Zealand, Portugal (IICT), Russia (Ministry of Finance), Switzerland (SDC), the UK Government (UK Aid), the European Union, and carried out with technical support from the International Fund for Agricultural Development (IFAD).

Todd S. Rosenstock
Mariana C. Rufino
Klaus Butterbach-Bahl
Eva Wollenberg
Meryl Richards

Contents

Contributors

Alain Albrecht Institute of Research for Development (IRD), Montpellier, France

Piet J.A. van Asten International Institute of Tropical Agriculture (IITA), Kampala, Uganda

Clement Atzberger University of Natural Resources (BOKU), Vienna, Austria

Germán Baldi Instituto de Matemática Aplicada San Luis, Universidad Nacional de San Luis and Consejo Nacional de Ciencia y Tecnología (CONICET), San Luis, Argentina

Lenny van Bussel Wageningen University and Research Centre, Wageningen, Netherlands

Klaus Butterbach-Bahl International Livestock Research Institute (ILRI), Nairobi, Kenya

Karlsruhe Institute of Technology, Institute of Meteorology and Climate Research, Atmospheric Environmental Research (IMK-IFU), Garmisch-Partenkirchen, Germany

C. Chang Commonwealth Scientific and Industrial Research Organisation (CSIRO), Townsville, QLD, Australia

Ngonidzashe Chirinda International Center for Tropical Agriculture (CIAT), Cali, Colombia

Eugenio Díaz-Pinés Karlsruhe Institute of Technology, Institute of Meteorology and Climate Research, Atmospheric Environmental Research (IMK-IFU), Garmisch-Partenkirchen, Germany

Ken E. Giller Plant Production Systems Group, Wageningen University, Wageningen, Netherlands

John P. Goopy International Livestock Research Institute (ILRI), Nairobi, Kenya

Jonathan Hickman Earth Institute, Columbia University, New York, USA

M.L. Jat International Maize and Wheat Improvement Centre (CIMMYT), New Delhi, India

R.K. Jat International Maize and Wheat Improvement Centre (CIMMYT), New Delhi, India

Borlaug Institute of South Asia, Pusa, Bihar, India

P. Kapoor International Maize and Wheat Improvement Centre (CIMMYT), New Delhi, India

Sean P. Kearney University of British Colombia, Vancouver, BC, Canada

Charlotte J. Klapwijk Plant Production Systems Group, Wageningen University and Research Centre, Wageningen, Netherlands

International Institute of Tropical Agriculture (IITA), Kampala, Uganda

Shem Kuyah World Agroforestry Centre (ICRAF), Nairobi, Kenya

Jomo Kenyatta University of Agriculture and Technology (JKUAT), Nairobi, Kenya

Cheikh Mbow World Agroforestry Centre (ICRAF), Nairobi, Kenya

Meine van Noordwijk World Agroforestry Centre (ICRAF), Bogor, Indonesia

David Pelster International Livestock Research Institute (ILRI), Nairobi, Kenya

Pytrik Reidsma Wageningen University and Research Centre, Wageningen, Netherlands

Meryl Richards Gund Institute for Ecological Economics, University of Vermont, Burlington, VT, USA

CGIAR Research Program on Climate Change, Agriculture, and Food Security (CCAFS)

Todd S. Rosenstock World Agroforestry Centre (ICRAF), Nairobi, Kenya

Mariana C. Rufino Center for International Forestry Research (CIFOR), Nairobi, Kenya

Gustavo Saiz Karlsruhe Institute of Technology, Institute of Meteorology and Climate Research, Atmospheric Environmental Research (IMK-IFU), Garmisch-Partenkirchen, Germany

Björn Ole Sander International Rice Research Institute (IRRI), Los Baños, Philippines

Tek B. Sapkota International Maize and Wheat Improvement Centre (CIMMYT), New Delhi, India

Gudeta W. Sileshi Freelance Consultant, Kalundu, Lusaka, Zambia

Sean M. Smukler University of British Colombia, Vancouver, BC, Canada

David Stern Maseno University, Maseno, Kenya

Clare Stirling International Maize and Wheat Improvement Centre (CIMMYT), Wales, UK

Philip K. Thornton CGIAR Research Program on Climate Change, Agriculture and Food Security (CCAFS), Nairobi, Kenya

Nigel Tomkins Commonwealth Scientific and Industrial Research Organisation (CSIRO), Livestock Industries, Townsville, QLD, Australia

Katherine L. Tully University of Maryland, College Park, MD, USA

Mark T. van Wijk International Livestock Research Institute, Nairobi, Kenya

Eva Wollenberg Gund Institute for Ecological Economics, University of Vermont, Burlington, VT, USA

CGIAR Research Program on Climate Change, Agriculture, and Food Security (CCAFS)

Chapter 1
Introduction to the SAMPLES Approach

Todd S. Rosenstock, Björn Ole Sander, Klaus Butterbach-Bahl, Mariana C. Rufino, Jonathan Hickman, Clare Stirling, Meryl Richards, and Eva Wollenberg

Abstract This chapter explains the rationale for greenhouse gas emission estimation in tropical developing countries and why guidelines for smallholder farming systems are needed. It briefly highlights the innovations of the SAMPLES approach and explains how these advances fill a critical gap in the available quantification guidelines. The chapter concludes by describing how to use the guidelines.

1.1 Motivation for These Guidelines

Agriculture in tropical developing countries produces about 7–9 % of annual anthropogenic greenhouse gas (GHG) emissions and contributes to additional emissions through land-use change (Smith et al. 2014). At the same time, nearly 70 % of the

T.S. Rosenstock (✉)
World Agroforestry Centre (ICRAF),
UN Avenue-Gigiri, PO Box 30677-00100, Nairobi, Kenya
e-mail: t.rosenstock@cgiar.org

B.O. Sander
International Rice Research Institute (IRRI), Los Baños, Philippines

K. Butterbach-Bahl
International Livestock Research Institute (ILRI), Nairobi, Kenya

Karlsruhe Institute of Technology, Institute of Meteorology and Climate Research, Atmospheric Environmental Research (IMK-IFU), Garmisch-Partenkirchen, Germany

M.C. Rufino
Center for International Forestry Research (CIFOR), Nairobi, Kenya

J. Hickman
Earth Institute, Columbia University, New York, USA

C. Stirling
International Maize and Wheat Improvement Centre (CIMMYT), Wales, UK

M. Richards • E. Wollenberg
Gund Institute for Ecological Economics, University of Vermont, Burlington, Vermont, USA

CGIAR Research Program on Climate Change, Agriculture, and Food Security (CCAFS)

T.S. Rosenstock et al. (eds.), *Methods for Measuring Greenhouse Gas Balances and Evaluating Mitigation Options in Smallholder Agriculture*,
DOI 10.1007/978-3-319-29794-1_1

technical mitigation potential in the agricultural sector occurs in these countries (Smith et al. 2008). Enabling farmers in tropical developing countries to manage agriculture to reduce GHG emissions intensity (emissions per unit product) is consequently an important option for mitigating future atmospheric GHG concentrations.

Our current ability to quantify GHG emissions and mitigation from agriculture in tropical developing countries is remarkably limited (Rosenstock et al. 2013). Empirical measurement is expensive and therefore limited to small areas. Emissions can be estimated for large areas with a combination of field measurement, modeling and remote sensing, but even simple data about the extent of activities is often not available and models require calibration and validation (Olander et al 2014). These guidelines focus on how to produce field measurements as a method for consistent, robust empirical data and to produce better models.

For all but a few crops and systems, there are no measured data for the emissions of current practices or the practices that would potentially reduce net emissions. For crops, significant information has been gathered for irrigated rice systems e.g., in the Philippines, Thailand, and China (Linquist et al. 2012; Siopongo et al. 2014) and for nitrous oxide emissions from China where high levels of fertilizer are applied (Ding et al. 2007; Vitousek et al. 2009). Yet measurements of methane from livestock—a major source of agricultural GHG emissions in most of the developing world—are lacking (Dickhöfer et al. 2014). Similarly, little to no information exists for most other GHG sources and sinks. Smallholder farms comprise a significant proportion of agriculture in the developing world in aggregate, as high as 98 % of the agricultural land area in China, for example, yet tend to escape attention as a source of significant emissions because of the small size of individual farms.

The dearth of empirical data contributes to why most tropical developing countries, all of which are non-Annex 1 countries of the UNFCCC, report emissions to the UNFCCC using Tier 1 methodologies with default emission factors, rather than more precise Tier 2 or Tier 3 methods and country-specific emission factors (Ogle et al. 2014). However, Tier 1 default emission factors represent a global average of data derived primarily from research conducted in temperate climates for monocultures, which is very different from the complex agricultural systems and landscapes typical of smallholder farms in the tropics. Given our knowledge of the mechanisms driving emissions and sequestration (e.g., temperature, precipitation, primary productivity, soil types, microbial activity, substrate availability), there is reason to believe that these factors represent only a rough approximation of the true values for emissions (Milne et al. 2013).

Field measurement of GHG emissions in tropical developing countries is generally done using methods developed in temperate developed countries. However, multiple factors complicate measurement of agricultural GHG sources and sinks in non-Annex 1 countries and necessitate approaches specific to the conditions common in these countries, including heterogeneity of the landscape, the need for low-cost methods, and the need for improving farmers' livelihood and food security.

Heterogeneous landscapes. Annex-1 countries are dominated by industrial agriculture, usually monocultures with commonly defined practices, over relatively large expanses. The combination of high research intensity and large-scale agriculture in developed countries creates a homogenous, relatively data-rich environment where point measurements of key sources (e.g., soil emissions from corn production in the Midwestern US or methane production from Danish dairy animals) can be extrapolated with acceptable levels of uncertainty to larger areas using empirical and process-based models (Del Grosso et al. 2008; Millar et al. 2010).

In contrast, many farmers (particularly smallholders) in tropical developing countries operate diversified farms with multiple crops and livestock, with field sizes often less than 2 hectares. For example, in western Kenya maize is often intercropped with beans, trees, or both and in regions with two rainy seasons, maize might be followed in the rotation by sorghum or other crops. Exceptions exist of course, such as in Brazil, where industrial farming is well established and farms can be thousands of hectares. Where heterogeneity does exist, it complicates the design of the sampling approach in terms of identifying the boundary of the measurement effort, stratifying the farm or landscape, and determining the necessary sampling effort. Capturing the heterogeneity of such systems, as well as comparing the effects of mitigation practices or agronomic interventions to improve productivity, often demands an impractical number of samples (Milne et al. 2013). Methods are needed to stratify complex landscapes and target measurements to the most important land units in terms of emissions and/or mitigation potential.

Resource limitations. People and institutions undertaking GHG measurements have different objectives, tolerances for uncertainty, and resources. Cost of research is one of the major barriers faced by non-Annex 1 countries in moving to Tier 2 or Tier 3 quantification methods. Some methods require sampling equipment, laboratory analytical capacity, and expertise that is not available in many developing countries. Furthermore, different spatial scales (e.g., field, farm, or landscape) require different methods and approaches. The chapters in this volume guide the user in choosing from available methods, taking into account the user's objectives, resources and capacity.

Improving livelihood and food security as a primary concern. The importance of improving farmer's livelihoods and capacity to contribute to food security though improved productivity must be taken into account in mitigation decision-making and the research agenda supporting those decisions. Measuring GHG emissions per unit area is a standard practice for accounting purposes, but measuring emissions per unit yield allows tracking of the efficiency of GHG for the yield produced and informs agronomic practices (Linquist et al. 2012). This volume considers productivity in targeting measurements and sampling design, along with recommendations for cost-effective yield measurements.

Improved data on agricultural GHG emissions and mitigation potentials provides opportunities to decision-makers at all levels. First and foremost, it allows governments and development organizations to identify high production, low-emission development trajectories for the agriculture sector. With the suite of farm- and landscape-level management options for GHG mitigation and improved productivity available for just about any site-specific situation, there are numerous options to select from. Country- or region-specific data allows more accurate comparison of

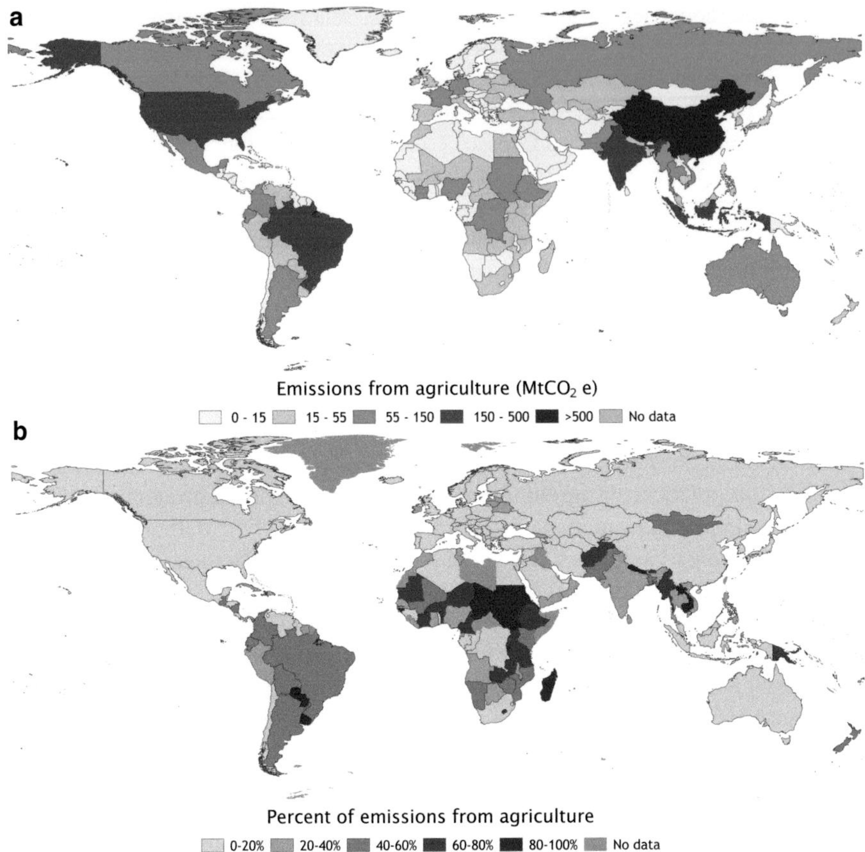

Fig. 1.1 (**a**) Total agricultural GHG emissions (GtCO$_2$e yr-1) by country (CH_4 and N_2O only). Data are average of emission figures from FAOSTAT database of GHG emissions from agriculture in 2010, EPA global emission estimates for 2010 and national reports to the United Nations Framework Convention on Climate Change (UNFCCC). If a country had not submitted a report to the UNFCC since the year 2000, we used only FAOSTAT and EPA data. (**b**) Percent of national emissions that come from agriculture, not including land-use, land-use change and forestry (LULUCF). Data from national reports to the UNFCCC

these options. Second, the prospects of the emerging green economy and potential for climate finance will dictate how emission reductions are both valued and verified. Verification, whether for Nationally Appropriate Mitigation Actions (NAMAs), Nationally Determined Contributions (NDCs), or product supply chain assessments, will require both reasonable estimates of baseline emissions and accurate quantification of emission reductions. Third, economies of tropical developing countries are largely dominated by agricultural production, and this sector contributes a significant fraction to their national GHG budgets (Fig. 1.1). Accurate data strengthen the basis for their negotiating position in global climate discussions.

1.2 Who Should Use These Guidelines?

These guidelines are intended to inform anyone conducting field measurements of agricultural greenhouse gas sources and sinks, especially to assess mitigation options in smallholder systems in tropical developing countries. The methods provide a standard for consistent, robust data that can be collected at reasonable cost with equipment often available in developing countries. They are also intended to provide end users of GHG data with a standard to evaluate methods used in previous efforts and inform future quantification efforts. The comparative analyses found in these chapters are accompanied by the recommended step-by-step instructions for the methods on the SAMPLES website (www.samples.ccafs.cgiar.org).

Potential users of the guidelines include:

- National agricultural research centers (NARS). NARS researchers can use these guidelines to establish protocols for greenhouse gas measurement from agriculture within their institution and ensure comparability with other research partners. They may also be used to review the robustness of existing measurement methods or for finding ways to reduce costs.
- Compilers of national GHG inventories. These guidelines are intended to provide methods for data collection to support the development of Tier 2 emission factors and the calibration of process-based models for Tier 3 approaches.
- Developers of national and subnational mitigation plans that include agriculture. Strategies to limit or reduce emissions take multiple forms: Low-Emission Development Strategies (LEDS), and Nationally Appropriate Mitigation Actions (NAMAs) and at the national scale, Nationally Determined Contributions (NDCs). Accurate information is required both in the planning phase, to establish baselines and compare potential interventions, and in the implementation phase, to measure, report, and verify (MRV) emissions reductions attributable to the strategy or policy. Field measurements are often necessary to generate national emission factors or calibrate models that can then be used in MRV systems. These guidelines should be used to ensure that field measurements methods are cost-effective, comparable across sites, and of sufficient accuracy.
- Agricultural commodity companies and agricultural development projects. These guidelines complement greenhouse gas accounting methodologies such as the Product Category Rules (PCRs) and carbon credit standards as well as agricultural greenhouse gas calculators such as EX-Ante Carbon Balance Tool (EX-ACT) (Bernoux et al. 2010) and Cool Farm Tool (Hillier et al. 2011). These methodologies and tools often require, or are improved by, user-input data corresponding to the project area, such as soil C stocks or emission factors for fertilizer application. These guidelines and the associated web resources provide methods—not usually covered in product and project standards—for the field measurements to generate these data.
- Students and instructors. Postgraduate students, advisors, and university instructors can use these guidelines as a manual in selecting research methods.

Box 1.1 Make Best Use of Limited Resources by Carefully Selecting Practices for Testing

GHG measurement is often undertaken with the purpose of comparing mitigation practices. Too often, those practices are chosen randomly or opportunistically, without explicit consideration of their feasibility or mitigation potential. The results of GHG measurement research will be more useful if practices for testing are identified in a systematic way with input from relevant decision-makers. This can be thought of as a process of "filtering" options from a laundry list of potentials to a few for further testing.

Identify the scope of practices for consideration

This can be seen as the "boundary" of potential options. Establishing a spatial boundary is a first step; this may be ecological (a watershed) or political (a county). Additionally, it is useful to further narrow the focus to particular agricultural activities or sectors. The criteria for doing so may include:

- Extent of an activity within the landscape. The targeting approach described by Rufino et al. (Chap. 2) is useful to determine this, as are agricultural census data and land-cover maps.
- Magnitude of emissions from a given agricultural activity. At the national scale, this can be estimated from FAOSTAT (FAOSTAT 2015), or the national communication to the UNFCCC. At farm or landscape scales, greenhouse gas calculators (Colomb et al. 2013) can provide a rough estimate.
- Stakeholder priorities. Government development plans and priorities may provide opportunities to incorporate mitigation practices that also improve production or livelihoods. Farmer unions and project funders may have priorities as well. It is good practice to consult a variety of stakeholders in identifying priority activities or sectors, including women and disadvantaged groups.
- Scale of practice changes to be considered. Different mitigation practices imply differing scales of change within an agricultural system. Some may be incremental practice changes (such as improved nitrogen-use efficiency), whereas others may modify the entire system (such as changing crops or animal breeds, or incorporating trees). Some mitigation options are not "practices" per se, but transformational changes such as different livelihoods or a change in land-use, such as changing from nomadic pastoralism to settled agriculture (Howden et al. 2011).

Identify potential practices

Once the geography and scope of the mitigation effort have been established, develop a list of practices that may be applicable. Ideas may come from interviews and surveys of stakeholder groups as well as published literature. The website accompanying this volume includes resources for this purpose.

Box 1.1 (continued) *Narrow the list of practices for testing*

Several criteria should be used to narrow the list of practices to a smaller feasible number for field-testing.

- *Likely mitigation potential.* While the purpose of field measurements is to provide accurate information on mitigation potential, expert judgment and currently available emission factors and models can allow a rough estimate to guide field measurements toward practices with the largest potential for reducing emissions. Again, some greenhouse gas calculators are useful for this purpose. The CGIAR Research Program on Climate Change, Agriculture, and Food Security is currently developing a tool specifically to rank the most effective mitigation practices in a given geographic area (Nayak et al. 2014).
- *Uncertainty of current information.* Sometimes, the most relevant mitigation practice may be one that is already well studied in the project area, or for which uncertainty around mitigation potential is generally low. In these cases, it may be better to focus field measurement efforts on practices for which uncertainty is high, or globally available emission factors are not relevant. If uncertainty has not been quantified, it may be valuable to conduct a small initial measurement effort and compare these results with outputs from available models. This can then guide the larger measurement campaign to areas most needed to reduce uncertainty.
- *Benefits for adaptation and livelihoods.* Reduction of greenhouse gas emissions is not the primary focus of farmers or, usually, policy makers. Practices should also be prioritized based on their benefits in terms of productivity, income, and resilience to climate change. Here, input from farmers and their organizations is critical. Likewise, there may be barriers to adoption that make a particular practice impractical or require supportive policies, such as high upfront investment or lack of access to markets (Wilkes et al. 2013).
- *Available resources.* Funding, labor, and time will necessarily limit the number of practices for which measurements can be conducted.

1.3 How to Use These Guidelines

The ten chapters in this volume are grouped into three categories that correspond with the steps necessary to conduct measurement (1) question definition, (2) data acquisition and (3) "option" identification (synthesis) (Fig. 1.2). Some readers, such as those looking to evaluate mitigation options for an agricultural NAMA, may want to go through each step. Readers interested in measurement methods for a particular GHG source can go directly to the associated chapter.

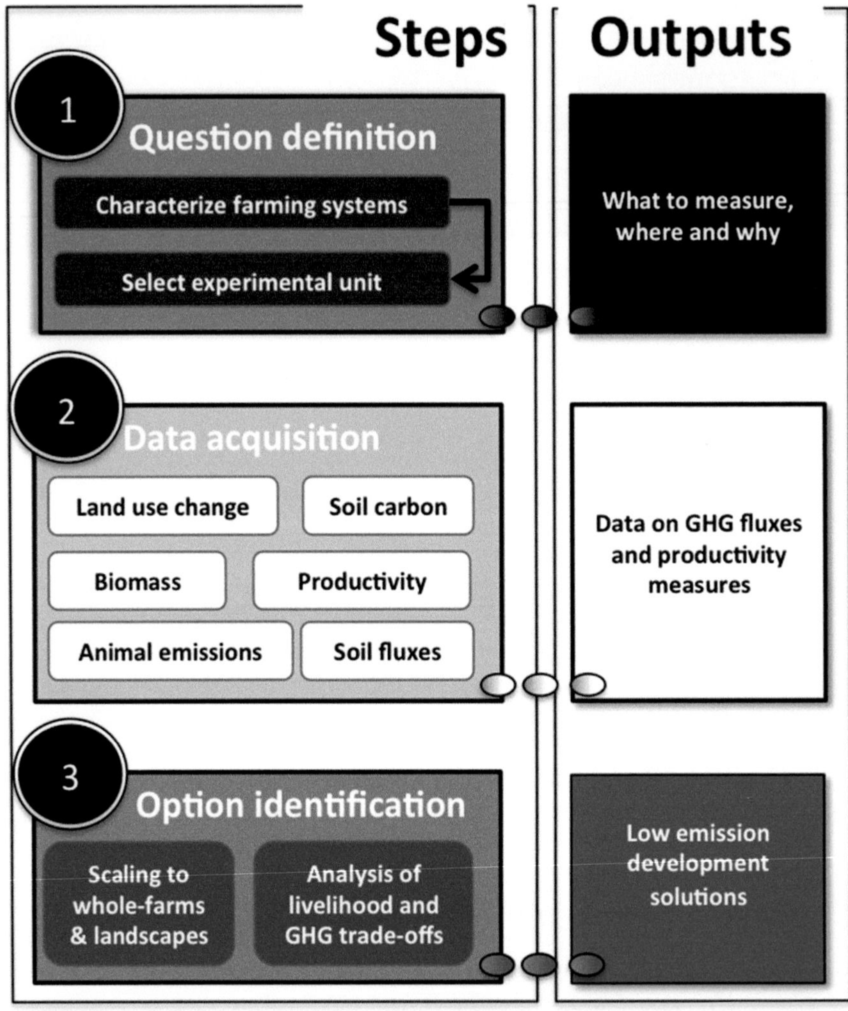

Fig. 1.2 Steps and their results of the SAMPLES approach. Each step yields inputs for subsequent steps, though components within each step are optional and subject to the interest of the inquiry.

Step 1. Question definition

Question definition defines the scope, boundaries and objectives of a measurement program. Measurement campaigns may be undertaken for a number of GHG quantification objectives such as developing emission factors, GHG inventories, or identifying mitigation options. The objective has considerable leverage on how and what is measured. In this volume, *Rufino et al.* (Chap. 2) describes methods for characterizing heterogeneous farming systems and landscapes, identifying the critical control points in terms of food security and GHG emissions in farming systems and landscapes. This characterization of the system generates fundamental information about the distribution and importance of farming activities in the landscape. Though often overlooked, depending on the preferences and priorities of donors or researchers, systems characterization is critical to target measurements to the most relevant areas in a landscape and stratify the landscape to inform sampling design.

Step 2. Data acquisition

Data acquisition is the "nuts and bolts" of quantification. It represents the activities that are conducted to measure and estimate GHG fluxes or changes in carbon stocks. The six chapters that make up this step discuss methods to quantify stocks, stock changes and fluxes of the major GHG sources and sinks including land-use and land-cover change (*Kearney and Smukler* Chap. 3), greenhouse gas emissions from soils (*Butterbach-Bahl et al.* Chap. 4), methane emissions due to enteric fermentation in ruminants (*Goopy et al.* Chap. 5), carbon in biomass (*Kuyah et al.* Chap. 6) and soil carbon stocks (*Saiz and Albrecht* Chap. 7). Methods to measure land productivity under agriculture—an essential input for tradeoff analysis—are treated separately (*Sapkota et al.* Chap. 8) (Table 1.1).

Each chapter provides a comparative analysis of existing methods for quantification, particularly evaluating methods across three key features—accuracy, scale, and cost (Table 1.2). Authors provide recommendations about how to select the optimal measurement approaches appropriate to the technical and financial constraints often encountered in developing countries, supplemented with discussion of the limitation of various methods. A central theme of the chapters is that GHG quantification is inherently inaccurate. The biogeochemistry of the processes that researchers are measuring coupled with the logistical practicalities of research mean that every measurement is only an estimate of the true flux. The researcher must therefore understand how different measurement approaches will affect their estimates and tailor measurement campaigns or quantification efforts to characterize the fluxes necessary to meet program objectives in a transparent and objective way. The resultant data on GHG fluxes produced from different sources and sinks can then be aggregated for partial or full GHG budgets using the guidelines from Chaps. 9–10.

Step 3. Estimation of emissions and analysis of mitigation options

The final step is to synthesize the results to identify emissions levels and mitigation options.

Data acquisition in Step 2 may take place at multiple scales, ranging from point measurements of individual farming activities (such as soil carbon measurements) to pixel analysis at various resolutions of land-use and land-cover change. It is then necessary to extrapolate these point measurements of individual features back to scales of interest (fields, farms, or landscapes). *Rosenstock et al.* (Chap. 9) describe the three principal ways that this can be accomplished: empirical, process-based models or a combination of both. *Van Wijk et al.* (Chap. 10) provide guidance on approaches to synthesize all the data to produce esti-

Table 1.1 Chapters of this volume and their associated IPCC source and sink categories (IPCC 1996, 2006)

SAMPLES chapter	1996 IPCC guidelines	2006 IPCC guidelines
Chapter 3: Determining GHG emissions and removals associated with land-use and land-cover change	5 Land-use change and forestry	3B Land
Chapter 4: Measuring GHG emissions from managed and natural soils	4C Rice cultivation	3C2 Liming
	4D Agricultural soils	3C3 Urea application
		3C4 Direct N_2O emissions from managed soils
		3C7 Rice cultivations
Chapter 5: Measuring methane emissions from ruminants	4A Enteric fermentation	3A1 Enteric fermentation
Chapter 6: Quantifying tree biomass carbon stocks and fluxes in agricultural landscapes	5A Changes in forest and other woody biomass stocks	3B1 Forest land
	5B Forest and grassland conversion	3B2 Cropland
	5C Abandonment of managed lands	3B3 Grassland
	5-FL Forest land	
	5-CL Cropland	
	5-GL Grassland	
Chapter 7: Methods for quantification of soil carbon stocks and changes	5B Forest and grassland conversion	3B2 Cropland
	5C Abandonment of managed lands	3B3 Grassland
	5D CO_2 emissions and removals from soil	
	5-FL Forest land	
	5-CL Cropland	
	5-GL Grassland	
Chapter 8: Yield estimation of food and non-food crops in smallholder production systems	4F Field burning of agricultural residues (for calculating residue quantities)	3C1b Biomass burning on croplands

mates of tradeoffs or synergies in various farm or landscape management activities-for example, activities that support mitigation as well as adaptation to climate change. Tradeoff analysis, though originating in the 1970s, has been developing rapidly due to increase in computing power and advances in theory and modeling frameworks. However, the authors stress that practical analysis has to include stakeholders to integrate their own perspectives and preferences for the analysis to be practically valuable. By developing estimates of GHG fluxes at relevant scales and analyzing tradeoffs, the approaches detailed in this volume can inform low-emissions development planning.

Table 1.2 Examples of measurements options and their accuracy, cost, and scale implications based on analyses in this volume

Method	Experimental considerations			Select uses
	Accuracy	Scale	Costs	
Enteric fermentation				
Empirical equations	Low, subject to variability in feed intake and emissions relationships	Large, many animals, herds, and inventories	Low, when based on just numbers of animals but increase when feed intake is measured	– Inventories
Respiration chambers	High temporal resolution measurements with sophisticated equipment	Small, limited to only a few animals	High, specialized equipment for accurate high resolution measurements and animal maintenance	– Emission factors – Mitigation options
SF6	Moderate to high	Small, animals and herds	Moderate, requires specialized equipment and skills	– Emission factors, especially of grazing animals – Mitigation options
Soil emissions				
Laboratory incubations	Low, measure emission potential and may not match field conditions	Large, with potential for many hundreds of samples that can span large spatial extents	(Relatively) low per sample due to minimal field requirements	– Emission potential – Identify hotspots of emissions – Mechanistic research – Model parameterization
Manual static chambers	Moderate, high spatial and temporal variability can lead to poor estimates	Moderate, with pooling methods capable of collecting data from many sites	Moderate, relatively cheap but field and lab costs become prohibitively expensive in many developing countries	– Inventories – Emission factors – Mitigation options
Automatic chambers	High, overcome temporal variability issues but limited in numbers because of costs	Small, generally only one site is measured at a time	High, the infield system represents a significant cost per measurement	– Emission factors – Mechanistic research

References

Bernoux M, Branca G, Carro A, Lipper L (2010) Ex-ante greenhouse gas balance of agriculture and forestry development programs. Sci Agric 67(1):31–40

Colomb V, Touchemoulin O, Bockel L, Chotte J-L, Martin S, Tinlot M, Bernoux M (2013) Selection of appropriate calculators for landscape-scale greenhouse gas assessment for agriculture and forestry. Environ Res Lett 8:015029

Del Grosso SJ, Wirth J, Ogle SM, Parton WJ (2008) Estimating agricultural nitrous oxide emissions. Trans Am Geophys Union 89:529–530

Dickhöfer U, Butterbach-Bahl K, Pelster D (2014) What is needed for reducing the greenhouse gas footprint? Rural21 48:31–33

Ding W, Cai Y, Cai Z, Yagi K, Zheng X (2007) Soil respiration under maize crops: effects of water, temperature, and nitrogen fertilization. Soil Sci Soc Am J 71(3):944–951

FAO (2015) FAOSTAT. Food and Agriculture Organization of the United Nations, Rome, Italy. http://faostat.fao.org. Accessed 10 April 2015

Hillier J, Walter C, Malin D, Garcia-Suarez T, Mila-i-Canals L, Smith P (2011) A farm-focused calculator for emissions from crop and livestock production. Environmental Modell Softw 26(9): 1070–1078

Howden SM, Soussana J-F, Tubiello FN, Chhetri N, Dunlop M, Meinke H (2011) Adapting agriculture to climate change. In: Cleugh H, Smith MS, Battaglia M, Graham P (eds) Climate change: science and solutions for Australia. CSIRO, Melbourne, pp 85–96

IPCC (1996) Revised 1996 IPCC Guidelines for National Greenhouse Gas Inventories. OECD, Paris

IPCC (2006) 2006 IPCC Guidelines for National Greenhouse Gas Inventories. Eggleston HS, Buendia L, Miwa K, Ngara T, Tanabe K (eds) Prepared by the National Greenhouse Gas Inventories Programme. IGES, Japan

Linquist B, Van Groenigen KJ, Adviento-Borbe MA, Pittelkow C, Van Kessel C (2012) An agronomic assessment of greenhouse gas emissions from major cereal crops. Glob Chang Biol 18:194–209

Millar N, Philip Robertson G, Grace PR, Gehl RJ, Hoben JP (2010) Nitrogen fertilizer management for nitrous oxide (N_2O) mitigation in intensive corn (Maize) production: an emissions reduction protocol for US Midwest agriculture. Insectes Soc 57:185–204

Milne E, Neufeldt H, Rosenstock T, Smalligan M, Cerri CE, Malin D, Easter M, Bernoux M, Ogle S, Casarim F, Pearson T, Bird DN, Steglich E, Ostwald M, Denef K, Paustian K (2013) Methods for the quantification of GHG emissions at the landscape level for developing countries in smallholder contexts. Environ Res Lett 8:015019

Nayak D, Hillier J, Feliciano D, Vetter S. CCAFS-MOT: a screening tool. Presented at a learning session during the 20th session of the Conference of the Parties to the United Nations Framework Convention on Climate Change, Lima, Peru, 3 Dec 2014. http://ccafs.cgiar.org/mitigation-options-tool-agriculture. Accessed 10 April 2015

Ogle SM, Olander L, Wollenberg L, Rosenstock T, Tubiello F, Paustian K, Buendia L, Nihart A, Smith P (2014) Reducing greenhouse gas emissions and adapting agricultural management for climate change in developing countries: providing the basis for action. Glob Chang Biol 20(1):1–6

Olander LP, Wollenberg E, Tubiello FN, Herold M (2014) Synthesis and review: advancing agricultural greenhouse gas quantification. Environ Res Lett 9:075003

Rosenstock TS, Rufino MC, Wollenberg E (2013) Toward a protocol for quantifying the greenhouse gas balance and identifying mitigation options in smallholder farming systems. Environ Res Lett 8, 021003

Siopongco JDLC, Wassmann R, Sander BO (2013) Alternate wetting and drying in Philippine rice production: feasibility study for a Clean Development Mechanism. IRRI Technical Bulletin No. 17. International Rice Research Institute, Los Baños Philippines, 14p

Smith P, Martino D, Cai Z, Gwary D, Janzen H, Kumar P, McCarl B, Ogle S, O'Mara F, Rice C, Scholes B, Sirotenko O, Howden M, McAllister T, Pan G, Romanenkov V, Schneider U, Towprayoon S, Wattenbach M, Smith J (2008) Greenhouse gas mitigation in agriculture. Philos Trans R Soc Lond B Biol Sci 363:789–813

Smith P, Bustamante M, Ahammad H, Clark H, Dong H, Elsiddig EA, Haberl H, Harper R, House J, Jafari M, Masera O, Mbow C, Ravindranath NH, Rice CW, Robledo Abad C, Romanovskaya A, Sperling F, Tubiello F (2014) Agriculture, Forestry and Other Land Use (AFOLU). In: Climate Change 2014: Mitigation of Climate Change. Contribution of Working Group III to the Fifth Assessment Report of the Intergovernmental Panel on Climate Change. Edenhofer O, Pichs-Madruga R, Sokona Y, Farahani E, Kadner S, Seyboth K, Adler A, Baum I, Brunner S, Eickemeier P, Kriemann B, Savolainen J, Schlömer S, von Stechow C, Zwickel T, Minx JC (eds.). Cambridge University Press, Cambridge, United Kingdom and New York, NY, USA.

Vermeulen SJ, Campbell BM, Ingram JSII (2012) Climate change and food systems. Annu Rev Environ Resour 37:195–222

Vitousek P, Naylor R, Crews T (2009) Nutrient imbalances in agricultural development. Science 324:1519–1520

Wilkes A, Tennigkeit T, Solymosi K (2013) National planning for GHG mitigation in agriculture : a guidance document. Mitigation of Climate Change in Agriculture Series 8. Food and Agriculture Organization of the United Nations, Rome, Italy. www.fao.org/docrep/018/i3324e/i3324e.pdf. Accessed 10 April 2015

Chapter 2
Targeting Landscapes to Identify Mitigation Options in Smallholder Agriculture

Mariana C. Rufino, Clement Atzberger, Germán Baldi, Klaus Butterbach-Bahl, Todd S. Rosenstock, and David Stern

Abstract This chapter presents a method for targeting landscapes with the objective of assessing mitigation options for smallholder agriculture. It presents alternatives in terms of the degree of detail and complexity of the analysis, to match the requirement of research and development initiatives. We address heterogeneity in land-use decisions that is linked to the agroecological characteristics of the landscape and to the social and economic profiles of the land users. We believe that as projects implement this approach, and more data become available, the method will be refined to reduce costs and increase the efficiency and effectiveness of mitigation in smallholder agriculture. The approach is based on the assumption that landscape classifications reflect differences in land productivity and greenhouse gas (GHG) emissions, and can be used to scale up point or field-level measurements. At local level, the diversity of soils and land management can be meaningfully summarized using a suitable typology. Field types reflecting small-scale fertility gradients are correlated to land

M.C. Rufino (✉)
Centre for International Forestry Research Institute (CIFOR), PO Box 30677 Nairobi, Kenya
e-mail: m.rufino@cgiar.org

C. Atzberger
University of Natural Resources (BOKU), Peter Jordan Strasse 82, Vienna 1190, Austria

G. Baldi
Instituto de Matemática Aplicada San Luis, Universidad Nacional de San Luis and Consejo Nacional de Ciencia y Tecnología (CONICET),
Ejército de los Andes 950, D5700HHW, San Luis, Argentina

K. Butterbach-Bahl
International Livestock Research Institute (ILRI), PO Box 30709 Nairobi, Kenya

Karlsruhe Institute of Technology, Institute of Meteorology and Climate Research, Atmospheric Environmental Research (IMK-IFU),
Kreuzeckbahnstr. 19, Garmisch-Partenkirchen, Germany

T.S. Rosenstock
World Agroforestry Centre (ICRAF), PO Box 30677, Nairobi, Kenya

D. Stern
Maseno University, PO Box 333, Maseno, Kenya

T.S. Rosenstock et al. (eds.), *Methods for Measuring Greenhouse Gas Balances and Evaluating Mitigation Options in Smallholder Agriculture*,
DOI 10.1007/978-3-319-29794-1_2

quality, land productivity and quite likely to GHG emissions. A typology can be a useful tool to connect farmers' fields to landscape units because it represents the inherent quality of the land and human-induced changes, and connects the landscape to the existing socioeconomic profiles of smallholders. The method is explained using a smallholder system from western Kenya as an example.

2.1 Introduction

Little is known about the environmental impact of smallholder agriculture, especially its climate implications. The lack of data limits the capacity to plan for low-carbon development, the opportunities for smallholders to capitalize on carbon markets, and the ability of low-income countries to contribute to global climate negotiations. Most importantly for smallholders, available information has not been linked to the effects on their livelihoods. Many research initiatives aim to close this information gap and will eventually lead to the adoption of mitigation practices in smallholder agriculture. Technically feasible mitigation practices do not necessarily represent plausible options, which are desirable for farmers. A key goal of mitigation in smallholder agriculture is the long-term benefit to the farmers themselves, achieved either through improved practices or subsidized as part of a global emissions reduction market. This chapter focuses on targeting the measurement of greenhouse gas (GHG) emissions in smallholder systems, as it is expected that this will also correspond to the potential for social impact of mitigation. Here targeting means the process of selecting units of a landscape where scientists or project developers will estimate a number of parameters to assess mitigation potential of land-use practices. Systematic selection of measurement locations ensures that measurements can be scaled up to give meaningful information for implementing mitigation measures.

Analysis of smallholder agriculture is a challenge because farming takes place in fragmented and diverse landscapes. Various actors may wish to target mitigation actions in this environment, including national and subnational governments who want to meet mitigation goals; project implementers at all levels; communities that wish to access carbon financing; and the research community that wants to contribute meaningfully to climate change mitigation. Although the spatial resolution and coverage of the assessment differ across actors, all face two basic questions related to emissions: how much mitigation can be achieved and where.

The scientific community conducts biophysical research to estimate the potential of soils to sequester carbon, and to estimate emissions of non-CO_2 gases from agriculture, forestry, and other land uses (AFOLU). If estimates of emission reductions are not available, the success of mitigation actions will be unknown. This is mostly the case in projects proposed in low-income countries where information on emissions and carbon sequestration potential is nonexistent or patchy. Most commonly where interventions are proposed, landscapes are considered uniform and equally effective for the mitigation actions promoted.

Before implementing mitigation projects, all actors should examine the mitigation objectives and use a structured targeting top-down, bottom-up, or mixed-method approach. The scientific community should use the same principles to increase the effectiveness of mitigation research, allow for comparability, and fill knowledge gaps at critical stages. The targeting of mitigation research projects and the implementation of mitigation actions are typically framed in terms of mitigation potential. Such assessments are carried out at relatively large scale and provide a range of achievable objectives, but do not connect directly with land users' realities. This is often done at an academic level without on-the-ground consultations and ignoring socioeconomic barriers.

We propose a targeting method using varied sources to support the analysis including geographical information systems (GIS), remote sensing (RS), socioeconomic profiles, and biophysical drivers of GHG emissions. In summary, we introduce a cost-effective method for selecting representative fields and landscape units as a basis for estimating GHG emissions, soil carbon stocks, land productivity and economic benefits from cultivated soils and natural areas. The objective of this chapter is to guide scientists and practitioners in their decisions to estimate GHG emissions, and to identify mitigation options for smallholders at whole-farm and landscape levels. This is a new area of research that links mitigation science with development, landscape ecology, remote sensing, and economic and social sciences to understand the consequences of land-use decisions on the environment.

The proposed approach is based on the assumptions that:

1. A landscape can be practically described using GIS and RS techniques that explain either landscape features associated with land-use and/or vegetation structure and functioning. The resulting landscape classification therefore also reflects differences in land productivity and GHG emissions, and can be used to scale up point or field-level measurements.
2. At the local level, the diversity of soils and land management can be meaningfully summarized using a suitable typology. Field types reflecting small-scale soil fertility gradients are correlated with land quality, land productivity (Zingore et al. 2007; Tittonell et al. 2010) and quite likely GHG emissions. Land productivity includes physical values (e.g., expressed in biomass per unit of land) and economic goods (e.g., expressed in monetary value per unit of land).
3. A typology is a useful tool to connect farmers' fields to landscape units because it represents the inherent quality of the land and human-induced changes. It can also connect the landscape to the existing socioeconomic profiles of smallholders.

To test the method, we used a smallholder system from Western Kenya as an example.

2.2 Initial Steps

The targeting approach stratifies landscapes of different complexity into different classes, to identify units that provide estimates of emission reductions representing larger areas. Figure 2.1 shows how a complex landscape can be split—using a

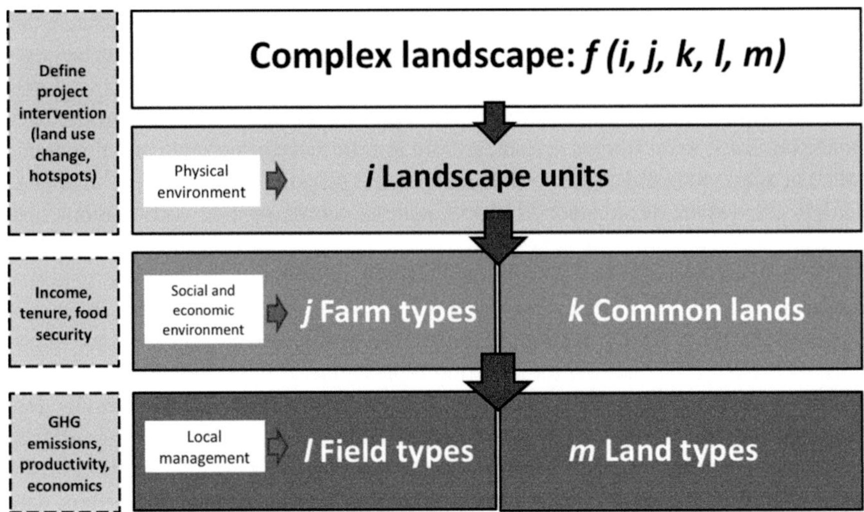

Fig. 2.1 Conceptual model of a nested targeting approach. The model indicates (*dashed boxes*) the sort of analyses conducted at each level

top-down approach—into smaller units (*i landscape units*) that have a common biophysical environment at regional scale. This disaggregation can be done using GIS and RS, assisted by existing secondary data. Landscape units can be further disaggregated into *j farm types* and *k common lands* to describe differences in the ways that individual households and communities access and use the land. The sort of units that link the land-to-land users will vary according to tenure systems in different territories, jurisdictions, and countries (Ostrom and Nagendra 2006). This step uses information on incomes, land tenure, and food security. It enables mitigation practices to be designed that are appropriate for heterogeneous rural communities, and where the land can be privately and communally managed. To make a connection with farming activities and ultimately with the level at which mitigation practices are implemented, farms and common lands can be disaggregated into *l field types* and *m land types*. This distinction may fade out in countries where the land is intensively used independently of the tenure system. The identified units can be studied in terms of land productivity, economic outputs, carbon stocks, GHG emissions, and the social and cultural importance of farming activities for rural families.

2.3 Top-Down Approach

We illustrate the steps to split a complex landscape (of any size) into homogeneous units using GIS and RS information and socioeconomic surveys to study mitigation potential (Fig. 2.1). This may be of interest, for example, where a carbon credit project is implemented, or if a district, province, or other authority wishes to assess the mitigation potential of a number of agricultural technologies. Once the landscape boundaries are defined, one can disaggregate the complex landscape into different units. If the landscape boundaries are not delineated, the analyst may choose to select an area that is representative of the larger region in order to extrapolate results. The landscape can be analyzed initially using a combination of RS and GIS. We suggest different approaches to disaggregate a landscape and decide where to conduct field measurements.

After selecting a landscape for assessment and developing a conceptual model of land-use and land-cover (LULC), the simplest method to identify landscape units is the exploration and visual interpretation of satellite imagery, preferably with the best available spatial resolution and observation conditions (e.g., peak of vegetation productivity). LULC classification (using object-based approaches and VHR imagery) and landscape classification (using RS vegetation productivity parameters) are more sophisticated methods of approaching a landscape. With visual interpretation, numerous landscape features can be characterized using physical (e.g., geomorphology, vegetation, disturbance signs) and human criteria (e.g., presence of population, land-use, and infrastructure). This yields relatively large, homogeneous landscape units (e.g., describing the mosaic of LULCs in an area). By comparison, automated LULC classification yields results at a much finer spatial scale. In most cases it maps the individual fields that make up a landscape. The process of automated LULC mapping involves:

1. Discriminating areas of general LULC types such as croplands or shrublands
2. Characterizing structural traits of all these types
3. Integrating areas and traits to identify homogeneous landscape units

The two first steps require the composition of the landscape to be characterized (i.e., the areas under each of the field or land types according to Fig. 2.1), and their spatial configuration (i.e., the arrangement of field or land types).

In landscapes with dominant smallholder agriculture, cultivated land can be easily recognized through the presence of regular plots with homogeneous surface brightness, and minor features such as ploughing or crop lines and infrastructure. In addition, the structural heterogeneity of cultivated areas can be assessed by the geometry of the fields (size and symmetry of the shapes), the presence of productive infrastructure and signs of disruption, such as woody encroachment within fields. Land under (semi-) natural vegetation can be characterized in terms of vegetation composition (share of trees, shrubs, and grass), signs of biomass removal or the presence of barren areas, and degradation (gullies, surface salt accumulation). Finally, in order to delimit landscape units, all descriptions should be integrated in a holistic manner using, for example, Gestalt-theory (Antrop and Van Eetvelde 2000) to identify and digitize potential discontinuities. This simple method has the potential to enhance the quality of broad-scale land-use studies, and can be performed using freely available imagery, like Google Earth, supported by online photographic archives such as “Panoramio” or “Confluence Project” (Ploton et al. 2012).

2.3.1 Landscape Stratification: An Example from East Africa

The Lower Nyando region of Western Kenya, which is dominated by smallholder producers, provides an example of the proposed approach. The CGIAR Program for Climate Change, Agriculture, and Food Security (CCAFS) promotes climate smart agriculture in this area. To develop and test our targeting approach, we used the three methods described above: (1) visual classification using VHR imagery, (2) LULC classification using object-based approaches and VHR imagery, and (3) landscape classification using medium to coarse resolution RS vegetation productivity parameters.

Visual Classification Using VHR Imagery

This is a quick and relatively inexpensive visual approach for exploring landscapes. The largest costs are the acquisition of the VHR images. Based on a QuickBird® image from the dry season (1 December 2008), six landscape classes were identified (Table 2.1 and Fig. 2.2). This initial classification can be used to test whether the units are indeed related to soil emissions and mitigation potential. The landscape classification is expected to reflect differences in land productivity and GHG emissions, because it captures inherent soil and vegetation variability.

Class delimitation criteria and mitigation opportunities are listed for each class in Table 2.1. The limits between the classes are determined by spatial changes in the detailed criteria. As expected, these changes can be abrupt or gradual, and the ability or experience of the mapper could lead to variable results.

The visual delineation may or may not coincide with regional biophysical gradients, as shown by a quick assessment of the topography of Nyando (Fig. 2.3). In our case study, the highlands coincided with areas allocated to cash crops, while the lowlands included a continuum from subsistence crops to wooded natural land types. Delineating a landscape on the sole basis of topography may be inaccurate and/or incomplete, yet the use of a digital elevation model (DEM) is an inexpensive option to simplify landscapes.

Land-Use and Land-Cover Classification Using Object-Based Approaches and VHR Imagery

The fine-scale analysis of actual LULC allows the interface between biophysical and human-induced processes to be captured. The automated methods are more complex than the visual interpretation described previously and require digital processing of remote sensing imagery. VHR satellite imagery with pixel resolution <1 m can be used for semiautomatic (supervised) mapping of LULC in heterogeneous and fine-structured landscapes with sparse vegetation cover. To make optimal use of the rich information provided by the VHR data, object-based approaches are recommended.

Table 2.1 List of visual classes determined for the Nyando study region, Kenya

	Class	Delimitation criteria	Mitigation opportunities
A	Cultivated land dominated by cash crops	Presence of an agricultural matrix, i.e., extensive (>70 % of the area) and connected (few identifiable large patches) cover. Most plots (>75 %) are comparatively large and of similar size (~1 ha), regular-shaped (rectangular), and have a heterogeneous color and brightness. Heterogeneity in this class originates from plough or crop lines, pointing to a crop cover. Presence of infrastructure (e.g., houses, storage places, etc.). No degradation signs (e.g., surface salt accumulation, lack of vegetation, gullies)	Agroforestry, fertilizer management
B	Natural vegetation	Presence of a matrix of any original vegetation type (forests, shrublands, savannahs). Trees or large shrubs are clearly distinguishable by their round shape or shadows in the images	Halting land and tree cover degradation
C	Mixed natural vegetation and agricultural land	No single cover type reaches 70 % of the area, and patches of crop, pasture, and natural vegetation are intermingled	Agroforestry, livestock management
D	Cultivated land dominated by subsistence crops	Same as A, but most plots are smaller, of variable area and shape (rounded, elongated, irregular). In this class, heterogeneity comes in addition from patches of herbaceous or shrubby vegetation within plots (a sign of land abandonment), and surface degradation	Fertilizer and manure management, agroforestry
E	Agricultural land dominated by grazing land	Same as A, but most plots are comparatively larger, have irregular shape (no bilateral symmetry), and lack of plough or crop lines. Frequent isolated trees or shrubs inside plots. Signs of infrastructure are less common than in A	Livestock management, manure management, agroforestry
F	Mixed cultivated land	Both elements of A and D are found intermingled within small areas	Agroforestry, fertilizer, and manure management

Compared to pixel-based approaches, object-based approaches permit the full exploitation of the rich textural information present in VHR imagery, as well as shape-related information. They also avoid "salt and pepper" effects when classifying individual pixels. Figure 2.4 summarizes the main steps of such an approach.

In a similar way to Fig. 2.2, the landscape is first segmented into small, homogeneous subunits or objects. This process is indicated in Fig. 2.4 as *image segmentation*. Input to this image segmentation is georectified, multilayered very high-resolution (VHR) satellite images. The resulting objects (also called "segments") are groups of adjacent pixels, which share similar spectral properties, and which are different from other pixels belonging to other objects.

To segment a landscape using VHR satellite images, the so-called segmentation algorithms are used. Contrary to the visual classification approach, objects/segments are

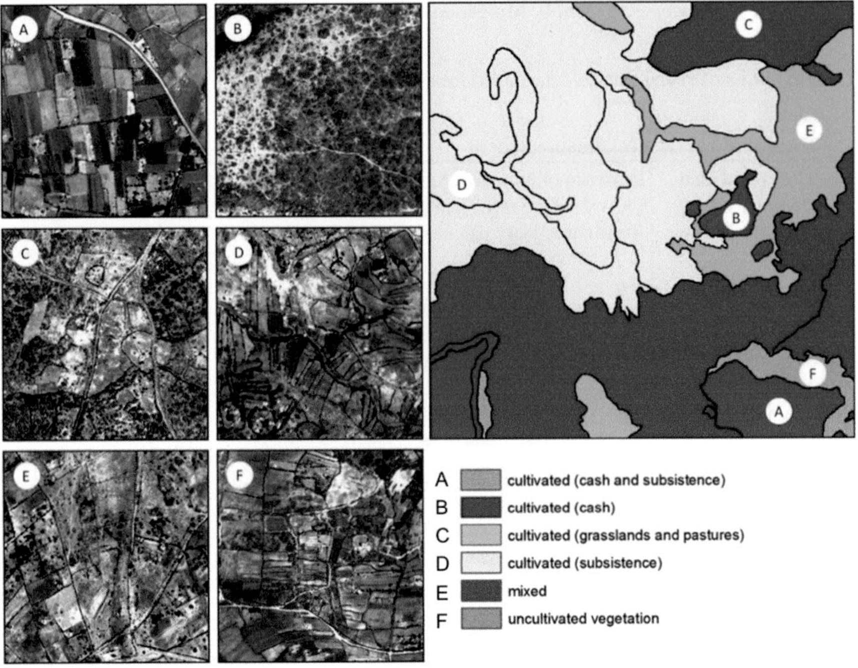

Fig. 2.2 Landscape analysis based on a visual inspection of landscape structure of Nyando, Western Kenya. (**a–f**) Are samples of the territory represented by the original QuickBird® image (all have the same spatial extent of 500 m). The larger panel on the *right* represents the six meaningful classes of landscape from the visual classification approach. Letters (A, B, C, D, E, and F) show the location of samples in the area (see explanations in Table 2.1)

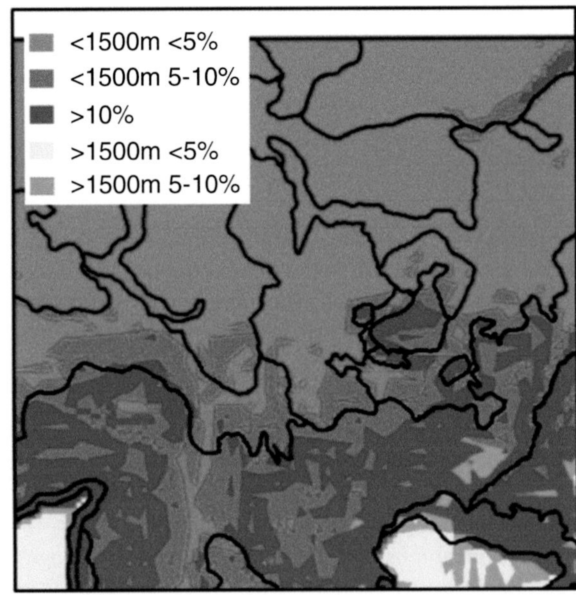

Fig. 2.3 Topographic characteristics of Nyando region. Altitude (masl) and slope (expressed as percentage) came from the Shuttle Radar Topography Mission (SRTM) digital elevation model (USGS 2004). The lines delineating the landscape units of Nyando are the same as in Fig. 2.2

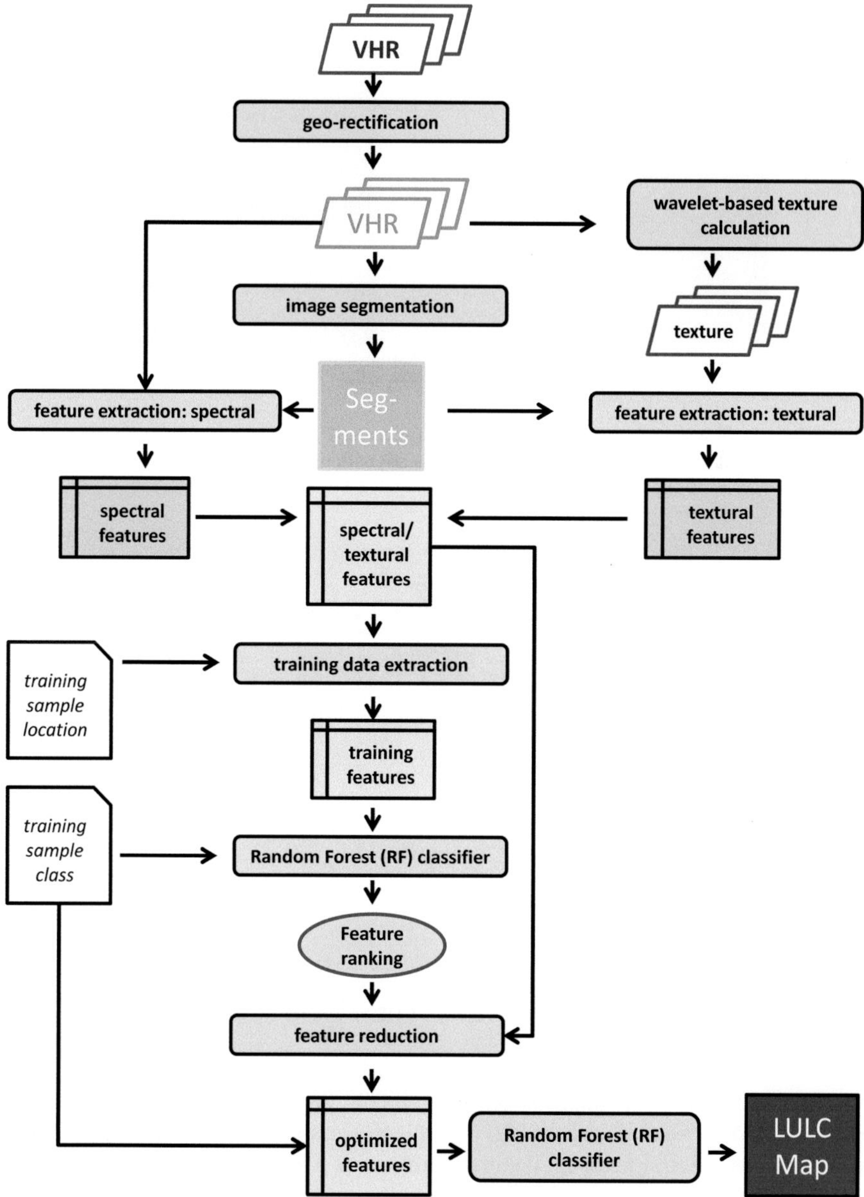

Fig. 2.4 Flowchart for object-based supervised classification of VHR imagery. The process yields a detailed LULC map of the area covered by the VHR satellite imagery, as well as information on the uncertainty of the classification outcome for each image object

identified in a fully automated manner. Both commercial and open source solutions exist for this task. Excellent open source solutions are, for example, QGIS (www.qgis.org/), GRASS GIS (grass.osgeo.org/) and ILWIS (www.ilwis.org/).

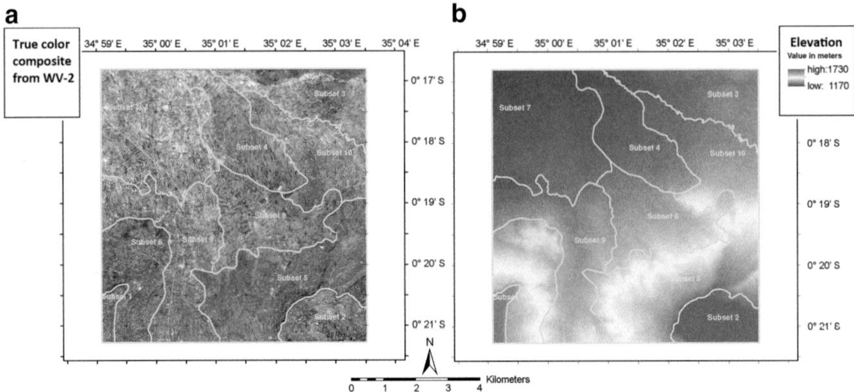

Fig. 2.5 Visualization of important steps of the supervised classification of the Nyando study region. (**a**) RGB image of WorldView-2® VHR imagery with manually delineated strata, (**b**) DEM of the region with strata

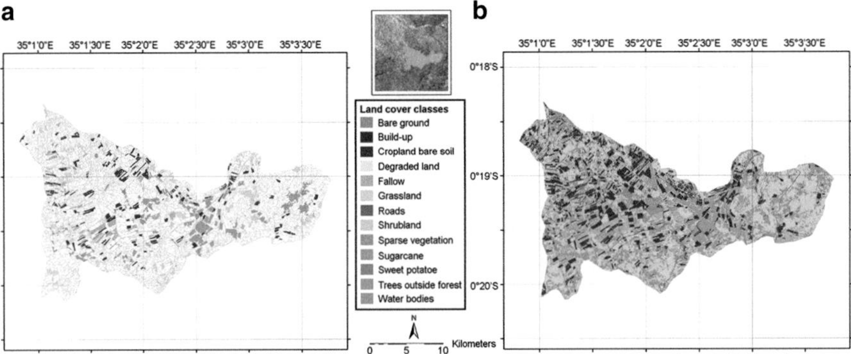

Fig. 2.6 (**a**) In situ information about the land-use/land-cover of training samples for one of the ten strata; the segmented image objects are also visible in *gray*, (**b**) classification result based on spectral and textural features of the WorldView-2® VHR image for the same stratum

After segmenting the image into image objects, an arbitrary number of features are extracted for each object. In Fig. 2.4, this process is labelled as *feature extraction*. Besides spectral features, textural features, as well as shape information, can be extracted. This information is used in a subsequent step to automatically assign each object to one of the user-defined LULC classes (process labelled as *Random (RF) forest classifier*). To "learn" the relationship between input features and class labels, training samples with known LULC must be provided in sufficient numbers and quality using a process called *training data extraction*.

Because the relation between input features and class label may change depending on image location (e.g., related to terrain and elevation), a stratified classification is recommended. For this task, before starting the classification process, the entire scene is (visually) split into a few (larger) regions (or strata) that can be considered homogeneous in terms of land-cover characteristics and the physical setting of the landscape.

The stratification is usually done just after the automated image segmentation (Fig. 2.4). Of course, results from other studies can be used as well (e.g., boundaries shown in Fig. 2.2). Figure 2.5a shows the RGB composite of a WorldView-2 image of the Nyando study area, and Fig. 2.5b, the corresponding DEM. In both maps, manually drawn landscape boundaries (strata) are also shown (yellow lines).

For one of the strata, Fig. 2.6a shows the available reference information obtained from fieldwork and complemented through visual image interpretation. These training samples are necessary for the RF classifier to "learn" the relationship between input features and class labels. The resulting object-based classification is shown for this landscape unit in Fig. 2.6b. The object limits (e.g., gray lines in Fig. 2.6a) have been automatically derived using GRASS GIS.

For the classification, several algorithms are available (e.g., maximum likelihood classifier, CART, kNN, etc.). Based on the authors' own and published experience, we exploited a widely used ensemble classifier called "random forest" (RF) which often yields good and robust classification results (Gislason et al. 2006; Rodriguez-Galiano et al. 2012; Toscani et al. 2013). RF uses bootstrap aggregation to create different training subsets, to produce a diversity of classification trees, each providing a unique classification result. For example, if 500 decision trees are grown inside the RF, one will obtain 500 class labels for each object. The final output class is obtained as the majority vote of the 500 individual labels (Breiman 2001). The proportion of votes of the winning class to the total number of trees used in the classification is a good measure of confidence; the higher the score, the more confident one can be that a class is correctly classified. Similarly, the margin calculated as the proportion of votes for the winning class minus the proportion of votes of the second class indicates how sure the classifier was in their decision. Such confidence indicators are not readily obtained using visual image interpretation. RF also produces an internal unbiased estimate of the generalization error, using the so-called "out-of-bag" (OOB) samples to provide a measure of the input features' importance through random permutation. Classification performance of the entire LULC map can be based on common statistical measures (overall accuracy (OA), producer's accuracy (PA) and user's accuracy (UA)) (Foody 2002) derived from the classification error matrix, using suitable validation samples. Figure 2.7 shows the resulting LULC map of Nyando obtained with this object-based classification approach and using VHR imagery from WorldView-2®.

Landscape Classification Using RS Vegetation Productivity Parameters

The two previous approaches are based on static descriptions of the landscape units (or of their constituent elements) in terms of LULC. However, alternative land traits can be explored to determine homogeneous landscape units. A promising alternative is the analysis of vegetation function in terms of the magnitude and temporal

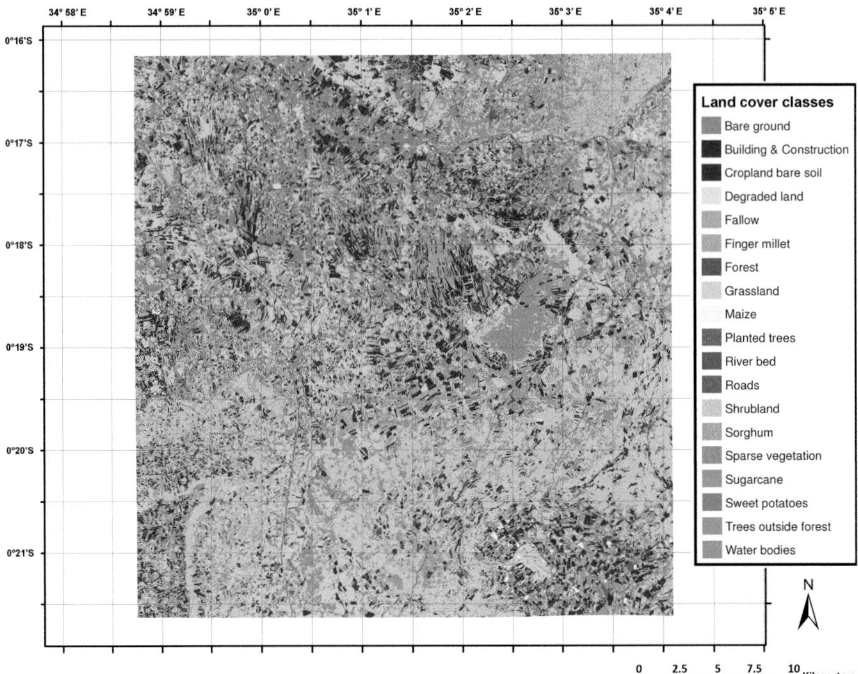

Fig. 2.7 LULC map of Nyando from WorldView-2® VHR imagery, using an object-based classification approach

variability of primary productivity (Paruelo et al. 2001). We tested this functional analysis in Lower Nyando, using the period 2000–2012. Vegetation primary productivity was assessed through the proxy variable Normalized Difference Vegetation Index (NDVI). This index has been of great value for biogeographical studies, allowing rough but widespread characterizations of the magnitude and temporal variability of productivity based on homogeneous measurements across wide spatial and temporal extensions and different ecosystems (Lloyd 1990; Xiao et al. 2004; Sims et al. 2006). In this example, we acquired NDVI data from the MODIS (Moderate Resolution Imaging Spectroradiometer) Terra instrument.[1] In this dataset, one image is produced every 16 days, leading to 23 images per year.

We selected from the 13-years × 23-dates database, only those values indicating good to excellent quality conditions (i.e., pixels not covered by clouds, and with a low to intermediate aerosol contamination). Then, we used the code TIMESAT v.3.1 to reconstruct temporal series (Jönsson and Eklundh 2002, 2004; Eklundh and Jönsson 2011). This tool fits smoothed model functions that capture one or two cycles of growth and decline per year. We selected an adaptive Savitzky-Golay

[1] Product coded as the MOD13Q1; spatial and temporal resolutions of 250 m and 16 days, respectively from the ORNL "MODIS Global Subsets: Data Subsetting and Visualization" online tool (http://daac.ornl.gov).

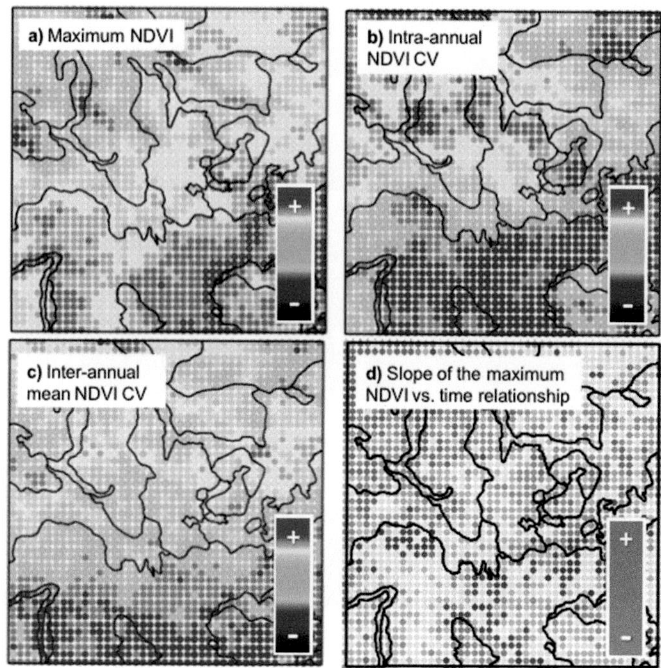

Fig. 2.8 Vegetation functioning depicting an average annual magnitude and seasonality, and interannual variability of primary productivity. (**a**) Maximum NDVI, (**b**) Intra-annual NDVI CV, (**c**) Interannual mean NDVI CV, (**d**) slope of the maximum NDVI versus time relationship. *Lines* represent homogeneous landscape units from the visual interpretation of Fig. 2.2

model (Jönsson and Eklundh 2002), assuming two vegetation growth cycles per year due to the natural bimodal behavior of rains in the study region. From the reconstructed temporal series (and by means of TIMESAT and the R v.2.15 statistical software), we calculated different functional metrics depicting average annual magnitude (e.g., mean, maximum NDVI) and seasonality (e.g., coefficient of variation (CV) of available values, number of growing seasons), and interannual variability (e.g., CV of mean annual values, annual trends) (Baldi et al. 2014).

For the sake of simplicity in the Lower Nyando example, Figure 2.8 presents: (a) NDVI maximum values as a proxy for carbon stocks of cultivated and uncultivated ecosystems; (b) intra-annual CV, describing whether the productivity is concentrated in a short period or distributed evenly through the year; (c) interannual CV of mean annual values, describing long-term productivity fluctuations; and (d) the slope of the maximum annual NDVI versus time relationship (Paruelo and Lauenroth 1998; Jobbágy et al. 2002).

Figure 2.9 shows the entire temporal range for the case of maximum annual values. Combined, structural and functional assessments provide essential information about the quality of the detected field or land types to study GHG mitigation potentials. Likewise, this approach may reveal functional divergences between a single

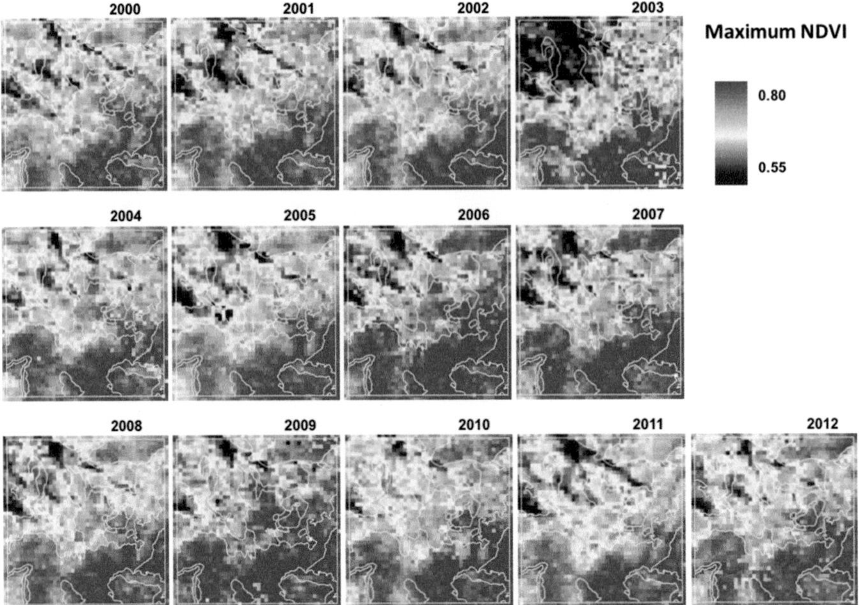

Fig. 2.9 Annual maximum NDVI value for the 2000–2012 period. *Lines* represent homogeneous landscape units from the visual interpretation of Fig. 2.2

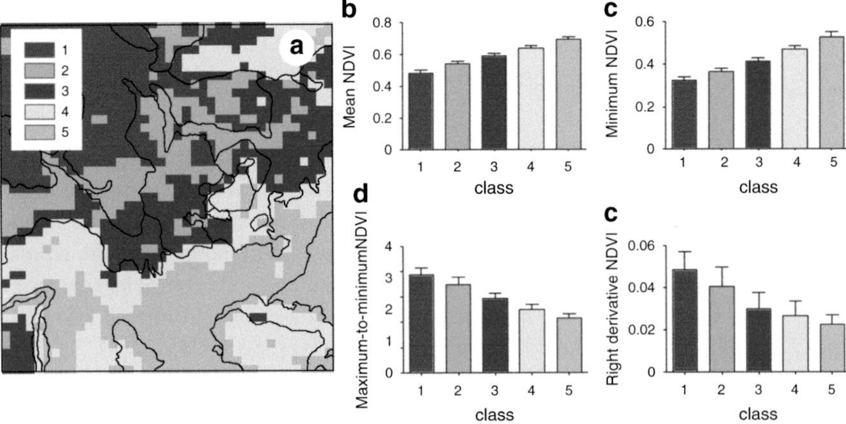

Fig. 2.10 (**a**) Nyando's five classes, based on unsupervised classification from four variables (**b–e**) exemplifying functional traits different from those presented in Figs. 2.7 and 2.8. Bars show averages and standard deviation for each class (depicted by the number and color). *Lines* in (**a**) represent homogeneous landscape units from the visual interpretation of Fig. 2.2

field or land type or convergences between different classes as shown in Figs. 2.8 and 2.9, with strong impacts on cascading ecosystem processes.

To identify landscape units using only functional information, we integrated functional attributes by applying an unsupervised classification procedure. In contrast with a LULC classification, we do not expect a priori conceptual scheme, both in terms of the number of classes and their identity. Functional classes often have to be split or merged to create a meaningful map, i.e., to show patterns of patches and corridors rather than isolated pixels ("salt and pepper" appearance). Using the unsupervised clustering algorithm ISODATA (Jensen 1996), we generated a map delimitating five different classes which reached our pattern-based expectations (Fig. 2.10). This approach revealed functional divergences between single farm types or common lands (e.g., western versus eastern cultivated areas dominated by cash crops), and convergences between different classes (e.g., western mixed shrubs and cultivated land versus eastern cultivated areas dominated by cash crops), with potential impacts on cascading ecosystem processes.

In addition to the landscape analysis, other on-the-ground information is needed for the development of a representative sampling design for smallholder systems before resource-consuming measurements of soil GHG fluxes or soil carbon and nitrogen stock inventories are implemented. The characterization of farmers' socioeconomic condition is important here, because this also affects resource management. On-farm variations in soil properties, which result from long-term differences in field management, create soil fertility gradients that may justify the use of a field typology.

2.4 Bottom-Up Approach

For some specific landscapes or agricultural systems there may be a wealth of field data that characterize the use of the land at field and farm level. This could include household surveys, soil surveys, productivity and economic assessments. This information comes at the price of laborious and costly data collection, and we encourage scientists and project developers to take advantage of existing field and farm data to inform the targeting of mitigation options at the local level. The analysis of these data informs the selection of field and farm types indicated in Fig. 2.1, which are the ultimate entry point for deciding where to carry out GHG measurements and identifying mitigation practices. This field-level characterization is especially useful in very fragmented landscapes, where topography, soils and long-term management create strong gradients in soil fertility and water retention capacity, which may lead to differences in emissions potential (Yao et al. 2010; Wu et al. 2010). We acknowledge that such detailed characterization may not be needed in simple landscapes with few land uses and relatively flat relief. Expert opinion by soil scientists can help decision-making about the location of field-level assessments.

We present a method that can be used to link the fields and farming practices with the landscape level, and emissions due to agricultural practices with potential for emissions reductions at scale. The method is based on assumptions 2 and 3 presented in Sect. 2.1: i.e., that the diversity of soils and land management can be meaningfully summarized using a field typology, which connects farmers' fields to landscape units representing inherent land quality and human-induced changes. There is evidence that field types can be defined on the basis of simple indicators that are correlated to land quality and land productivity. Research in Western Kenya and Zimbabwe shows the relationship between soil quality, intensity of management, and land productivity (Tittonell et al. 2005, 2010; Zingore et al. 2007), which we believe are correlated to soil GHG emissions.

A field typology can be derived a priori using information collected in household surveys. This can help connect field management with farm types, defined by livelihood indicators, including food and tenure security. Including these dimensions in the analysis provides an opportunity to link mitigation with food security and poverty to estimate trade-offs and synergies. Such an analysis permits an assessment of the feasibility of mitigation for different farmers and identification of the incentives needed for adoption. Land users can assess and weigh up the livelihood benefits of different practices (e.g., income, increased production of food) and the costs of implementing such practices.

Using the Lower Nyando site, we show how to use household and field surveys to support targeting at a local level and how to link it to the selected landscape. We collected existing information on households and farm management. The lower Nyando site was characterized using the IMPACTlite tool (Rufino et al. 2012a, b) that gathered generic data to analyze food security, adaptation, and mitigation in smallholder agriculture. A comprehensive household survey was conducted to characterize household structure, asset ownership, farm production, costs and benefits of farming activities, other sources of income generation, and food consumption (Rufino et al. 2012a, b). Using the farm household characterization, and to elaborate the field typology, fields recorded in the household survey were measured, georeferenced and additional management data were collected. The household survey covered three production systems across the sampling frame of the Kenyan CCAFS site of Nyando (Förch et al. 2013), and included 200 households. A field typology was built on the basis of field type scores collected through a survey (see forms in Appendix). A subsample of fields was selected randomly to represent the field types.

2.4.1 Field Typology Definition

The field typology must reflect inherent soil fertility resulting from soil type and long-term management. The process of defining the field typology is dependent on the landscape within which the project works and the sociocultural norms of the farmers. For example, crop diversity may be considered as a sign of productive land in subsistence agriculture systems. Adjusting the weighting to take into account local knowledge is important to link well with ground truths.

The scores obtained through this process are simply a tool for subdividing fields based on easily obtainable data, analogous to a rapid rural appraisal (Dorward et al. 2007). It is often justifiable to adjust the weightings based on the data, by identifying the common characteristics of the field types and checking that the subdivisions are indeed meaningful. Whenever possible the classification should be counter checked against the common sense evaluation of an experienced field officer on the ground.

At the Nyando site, we used a number of variables to define a field type score:

- *Crop*. This score is the sum of the crops that each household is cultivating in one plot. Intensively managed fields are cropped with several crops, which often receive more agricultural inputs than other fields.
- *Fertilizer use*. This score distinguishes organic and inorganic fertilizers. Manure was given a score of 2 and other inorganic fertilizers a score of 3. It was assumed that fields receiving inorganic fertilizers are managed more intensively than fields that only receive animal manure.
- *Number of subplots*. This is the number of subplots within a given field or plot. Subplots are units within a field or permanent land management structure that can change in space or time. This aims to capture the spatial and temporal allocation of land to crops, crop mixtures, and the combination of annual and perennial crops in intercropping, permanent and seasonal grazing land.
- *Location of field*. Fields next to the homestead receive a score of 2, while fields further away from the house receive a score of 0. This assumes that fields close to the homestead receive preferential land management (e.g., fertilization, addition of organic matter, weeding) compared to fields that are far away.
- *Signs of erosion*. Fields differing in visible signs of erosion obtained different scores, depending on severity. For gulley, rill and gulley, sheet, rill erosion, fields received a score of 0. Sheet erosion or no visible erosion obtained a score of 1.

Plots with a score higher than 10 were labelled as field type 1. Those with scores between 4 and 10 were labelled as field type 2, and those with scores lower than 4 were labelled as field type 3. The process of defining scores for each variable involved making judgments about correlations and data quality. The end scores were then investigated, definitions adjusted and natural cut-off points identified. The identification of natural cut-offs for the field types is a delicate process because the scoring tool is crude enough that one would not expect a substantial difference on the ground between borderline cases. A useful guideline is that borderline cases should not be either under- or overrepresented in any field type.

2.5 Combining Top-Down and Bottom-Up: The Basis for Scaling Up

The field typology sampled across households represents the diversity of land management practices. If it is combined with a land-use classification, it connects local management with landscape characteristics as indicated in Fig. 2.11. Provided that land-use units or land classes have been sampled at field-level, or that spatially explicit information is available on the diversity of field types, connecting these two layers may provide a measure of variability on GHG emissions, productivity, and livelihood

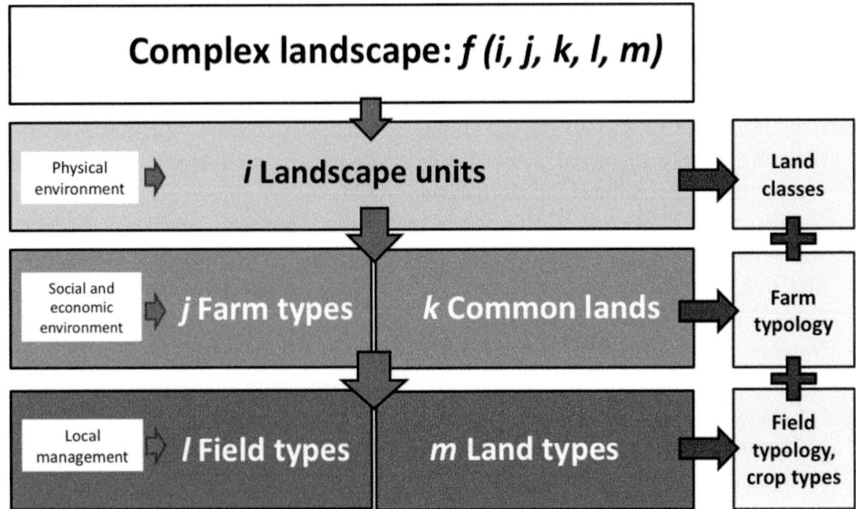

Fig. 2.11 Conceptual model and products of the nested targeting approach. The model indicates the sort of outputs obtained at each level. The integration of all level measurements conducted at field-level is to be scaled up

indicators. To achieve this, enough field sites have to be selected to represent each landscape class, and must be monitored for GHG emissions, carbon stock changes, production of biomass, and other variables of interest. The number of replicates or field sites to represent a landscape class will depend on within-class heterogeneity, and the resources available for monitoring emissions. An absolute minimum of three replicates per land class is required to estimate biophysical parameters.

The advantage of selecting replicated field sites that correspond to landscape classes is the possibility to scale up (i.e., to estimate project-level benefits and trade-offs with livelihood indicators). It also provides an opportunity to extrapolate findings to similar environments. In the case of lower Nyando, we combined the field typology derived from a household characterization with the landscape description including five classes or units shown in Fig. 2.10. "Landscape plots" were selected to represent field types using landscape units where we monitored GHG emissions, analyzed carbon stocks, and estimated productivity and the economics of production. We present here the results of 12 months of monitoring GHG emissions aggregated at field and landscape level (Fig. 2.12). The information provided a comprehensive database to estimate emissions potential and trade-offs with other socioeconomic indicators, such as income and land productivity. Additional field sites were added to compensate for areas poorly represented by the household survey and to include natural areas. This can be a serious disadvantage of using secondary data in a bottom-up approach, where householders neglect natural areas such as woodlands or wetlands during interviews. Natural areas were selected from the landscape analysis, where natural vegetation units were mapped.

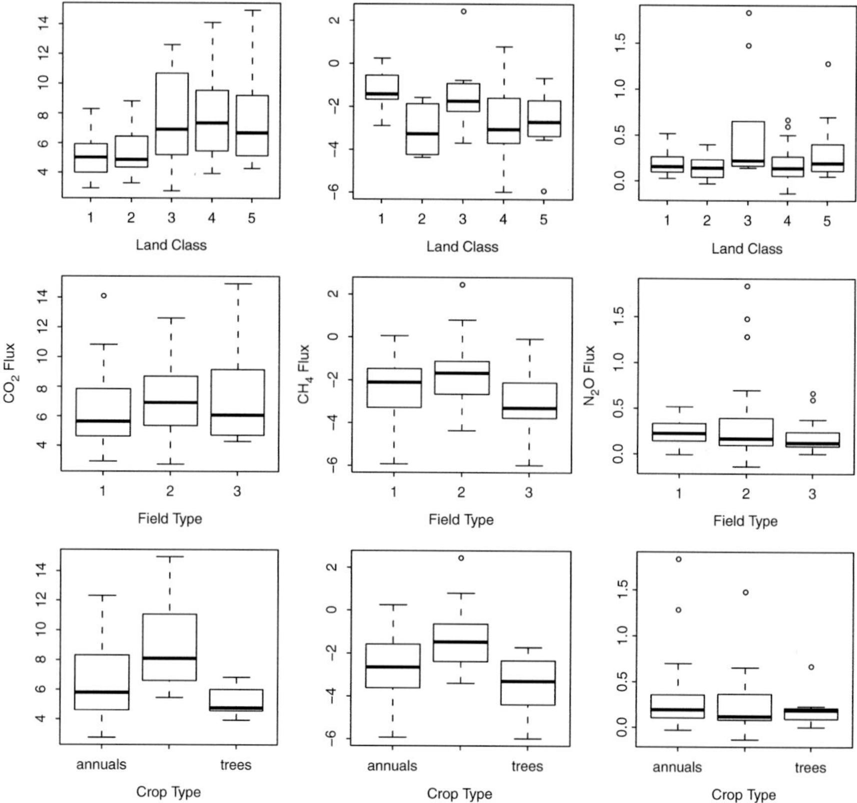

Fig. 2.12 Cumulative annual emissions of CO_2 (Mg C-CO_2 m^{-2} year^{-1}), CH_4 (kg C-CH_4 m^{-2} year^{-1}), and N_2O (kg N-N_2O m^{-2} year^{-1}) from 60 different fields located in Lower Nyando in Western Kenya split by land class, field type, crop type, and landscape position (Pelster et al. 2015).

2.6 Conclusions

A methodology is presented to target mitigation research at field, farm-, and landscape level. It uses both a top-down and a bottom-up approach to capture local diversity in soils and management practices, and landscape heterogeneity. It enables generic recommendations to be made about scaling up alternative mitigation options. The methods can fit the purposes of diverse projects, including the targeting of GHG measurement or the testing of carbon sequestration practices. The products generated such as land-use or land class maps and selected field types allow field sites to be selected for monitoring biophysical parameters. Once monitoring of GHG emissions, productivity, and economics are finalized, the nested approach suggested here provides a basis for scaling up, which can be achieved using different analytical methods discussed in Chap. 10 of this volume.

2.7 Appendix

Field typology survey

Date:
Surveyor:

HH ID: ____________________ Name of respondent: ____________________

PLOT LOCATION AND SIZE

South______________ East________________ Error________

	Plot	Subplot	Subplot	Subplot
ID				
Area (m2)				
Land cover				
Photo ID				

Land tenure:	Do animals graze the plot?	Does the farmer burn the plot?
Communal Rented Owned	regularly sometimes never	regularly sometimes never

Agricultural practices

Crops commonly planted in field

Crop (e.g. Maize) Highest yields (local units)

____________________ ____________________

____________________ ____________________

____________________ ____________________

Land cover prior to agriculture:

Forest

Grass or shrubland

unknown

How many years ago was it covered to agriculture (circle one):

0-2 2-5 5-10 >10 unknown

Are fertilizers applied?

Yes or No

If yes, which sub-plot?

____

YES, FERTILIZERS ARE APPLIED

Type	Amount	Crop

Type (eg)
UREA
CAN
MANURE
AMOUNT = PER PLOT
ID WHICH CROP

What is your best plot (or subplot) and why?

Woody cover (%)	Herbaceous cover (%):	Visible evidence of erosion
<4 4 - 15 15 - 40	<4 4 - 15 15-40	Rill Sheet Gully none
40 - 65 >65	40 - 65 >65	

References

Antrop M, Van Eetvelde V (2000) Holistic aspects of suburban landscapes: visual image interpretation and landscape metrics. Landsc Urban Plan 50:43–58

Baldi G, Houspanossian J, Murray F, Rosales AA, Rueda CV, Jobbágy EG (2014) Cultivating the dry forests of South America: diversity of land users and imprints on ecosystem functioning. J Arid Environ 123: 47–59 doi:10.1016/j.4

Breiman L (2001) Random forests. Mach Learn 45:5–32

Dorward P, Shepherd D, Galpin M (2007) Participatory farm management methods for analysis, decision making and communication. FAO, Rome, p 48

Eklundh L, Jönsson P (2011) Timesat 3.1 Software Manual. Lund University, Lund, Sweden

Foody GM (2002) Status of land cover classification accuracy assessment. Remote Sens Environ 80:185–201

Förch W, Kristjanson P, Thornton P, Kiplimo J (2013) Core sites in the CCAFS regions: Eastern Africa, West Africa and South Asia, Version 3. CGIAR Research Program on Climate Change, Agriculture and Food Security (CCAFS), Copenhagen, Denmark. http://ccafs.cgiar.org/initial-sites-ccafs-regions

Gislason PO, Benediktsson JA, Sveinsson JR (2006) Random forests for land cover. Pattern Recogn Lett 27:294–300

Jensen JR (1996) Introductory digital image processing: a remote sensing perspective. Pearson Prentice Hall, Upper Saddle River

Jobbágy EG, Sala OE, Paruelo JM (2002) Patterns and controls of primary production in the Patagonian steppe: a remote sensing approach. Ecology 83:307–319

Jönsson P, Eklundh L (2002) Seasonality extraction by function fitting to time-series of satellite sensor data. IEEE Trans Geosci Remote 40:1824–1832

Jönsson P, Eklundh L (2004) TIMESAT—a program for analyzing time-series of satellite sensor data. Comput Geosci 30:833–845

Lloyd D (1990) A phenological classification of terrestrial vegetation cover using shortwave vegetation index imagery. Int J Remote Sens 11:2269–2279

Ostrom E, Nagendra H (2006) Insights on linking forests, trees, and people from the air, on the ground, and in the laboratory. Proc Natl Acad Sci 103(51):19224–19231

Paruelo M, Lauenroth WK (1998) Interannual variability of NDVI and its relationship to climate for North American shrublands and grasslands. J Biogeogr 25:721–733

Paruelo JM, Jobbágy EG, Sala OE (2001) Current distribution of ecosystem functional types in temperate South America. Ecosystems 4:683–698

Pelster, DE, MC Rufino, TS Rosenstock, J Mango, G Saiz, E Diaz-Pines, G Baldi, K Butterbach-Bahl 2015 Smallholder African farms have very limited GHG emissions. Biogeosciences Discussions 12, 15301–15336

Ploton P, Pélissier R, Proisy C, Flavenot T, Barbier N, Rai SN, Couteron P (2012) Assessing aboveground tropical forest biomass using Google Earth canopy images. Ecol Appl 22:993–1003

Rodriguez-Galiano VF, Ghimire B, Rogan J, Chica-Olmo M, RigolSanchez JP (2012) An assessment of the effectiveness of a random forest classifier for land-cover classification. ISPRS J Photogramm Remote Sens 67:93–104

Rufino MC, Quiros C, Boureima M, Desta S, Douxchamps S, Herrero M, Kiplimo J, Lamissa D, Mango J, Moussa AS, Naab J, Ndour Y, Sayula G, Silvestri S, Singh D, Teufel N, Wanyama I (2012a) Developing generic tools for characterizing agricultural systems for climate and global change studies (IMPACTlite—phase 2). Report of Activities 2012. Submitted by ILRI to the CGIAR Research Program on Climate Change, Agriculture and Food Security (CCAFS), Copenhagen, Denmark

Rufino MC, Quiros C, Teufel N, Douxchamps S, Silvestri S, Mango J, Moussa AS, Herrero M (2012b) Household characterization survey—IMPACTlite training manual. Working document, CGIAR Research Program on Climate Change, Agriculture and Food Security (CCAFS), Copenhagen, Denmark

Sims DA, Rahman AF, Cordova VD, El-Masri BZ, Baldocchi DD, Flanagan LB, Goldstein AH, Hollinger DY, Misson L, Monson RK, Oechel WC, Schmid HP, Wofsy SC, Xu L (2006) On the use of MODIS EVI to assess gross primary productivity of North American ecosystems. J Geophys Res Biogeo 111. doi:10.10-9/2006JG000162

Tittonell P, Vanlauwe B, Leffelaar PA, Shepherd KD, Giller KE (2005) Exploring diversity in soil fertility management of smallholder farms in western Kenya: II. Within-farm variability in resource allocation, nutrient flows and soil fertility status. Agr Ecosyst Environ 110:166–184

Tittonell P, Muriuki A, Shepherd KD, Mugendi D, Kaizzi KC, Okeyo J, Verchot L, Coe R, Vanlauwe B (2010) The diversity of rural livelihoods and their influence on soil fertility in agricultural systems of East Africa—a typology of smallholder farms. Agr Syst 103:83–97

Toscani P, Immitzer M, Atzberger C (2013) Wavelet-based texture measures for object-based classification of aerial images. Photogramm Fernerkund Geoin 2:105–121

USGS (2004) Shuttle Radar Topography Mission, 1 Arc Second scene SRTM_u03_n008e004, Unfilled Unfinished 2.0, Global Land Cover Facility, University of Maryland, College Park, MD, February 2000

Wu X, Yao Z, Brüggemann N, Shen ZY, Wolf B, Dannenmann M, Zheng X, Butterbach-Bahl K (2010) Effects of soil moisture and temperature on CO_2 and CH_4 soil–atmosphere exchange of various land use/cover types in a semi-arid grassland in Inner Mongolia, China. Soil Biol Biochem 42:773–787

Xiao X, Zhang Q, Braswell B, Urbanski S, Boles S, Wofsy S, Berrien M, Ojima D (2004) Modeling gross primary production of temperate deciduous broadleaf forest using satellite images and climate data. Remote Sens Environ 91:256–270

Yao Z, Wu X, Wolf B, Dannenmann M, Butterbach-Bahl K, Brüggemann N, Chen W, Zheng X (2010) Soil-atmosphere exchange potential of NO and N_2O in different land use types of Inner Mongolia as affected by soil temperature, soil moisture, freeze-thaw, and drying-wetting events. J Geophys Res 115, D17116

Zingore S, Murwira HK, Delve RJ, Giller KE (2007) Soil type, historical management and current resource allocation: three dimensions regulating variability of maize yields and nutrient use efficiencies on African smallholder farms. Field Crop Res 101:296–305

Chapter 3
Determining Greenhouse Gas Emissions and Removals Associated with Land-Use and Land-Cover Change

Sean P. Kearney and Sean M. Smukler

Abstract This chapter reviews methods and considerations for quantifying greenhouse gas (GHG) emissions and removals associated with changes in land-use and land-cover (LULC) in the context of smallholder agriculture. LULC change contributes a sizeable portion of global anthropogenic GHG emissions, accounting for 12.5 % of carbon emissions from 1990 to 2010 (Biogeosciences 9:5125–5142, 2012). Yet quantifying emissions from LULC change remains one of the most uncertain components in carbon budgeting, particularly in landscapes dominated by smallholder agriculture (Mitig Adapt Strateg Glob Chang 12:1001–1026, 2007; Biogeosciences 9:5125–5142, 2012; Glob Chang Biol 18:2089–2101, 2012). Current LULC monitoring methodologies are not well-suited for the size of land holdings and the rapid transformations from one land-use to another typically found in smallholder landscapes. In this chapter we propose a suite of methods for estimating the net changes in GHG emissions that specifically address the conditions of smallholder agriculture. We present methods encompassing a range of resource requirements and accuracy, and the trade-offs between cost and accuracy are specifically discussed. The chapter begins with an introduction to existing protocols, standards, and international reporting guidelines and how they relate to quantifying, analyzing, and reporting GHG emissions and removals from LULC change. We introduce general considerations and methodologies specific to smallholder agricultural landscapes for generating activity data, linking it with GHG emission factors and assessing uncertainty. We then provide methodological options, additional considerations, and minimum datasets required to meet the varying levels of reporting accuracy, ranging from low-cost high-uncertainty to high-cost low-uncertainty approaches. Technical step-by-step details for suggested approaches can be found in the associated website.

S.P. Kearney • S.M. Smukler (✉)
University of British Colombia, Vancouver, BC, Canada
e-mail: sean.smukler@ubc.ca

T.S. Rosenstock et al. (eds.), *Methods for Measuring Greenhouse Gas Balances and Evaluating Mitigation Options in Smallholder Agriculture*,
DOI 10.1007/978-3-319-29794-1_3

3.1 Introduction

Land-use and land-cover (LULC) change contributes a sizeable portion of global anthropogenic GHG emissions, accounting for an estimated 12.5 % of carbon emissions from 1990 to 2010 (Houghton et al. 2012). Significant demographic and socio-economic pressures are exerted on carbon storing land uses such as forests in the tropics yet distribution and rates of change (e.g., loss of forests and agricultural intensification) in tropical smallholder landscapes remain very uncertain (Achard et al. 2002). Much of this uncertainty stems from the substantial heterogeneity of LULC that exists, often at very fine spatial scales, in such landscapes. Even within LULC categories, significant heterogeneity in carbon stocks often occurs as a result of drivers specific to smallholder agriculture, such as fallow rotations, uneven canopy age distribution, and integrated crop–livestock systems (Maniatis and Mollicone 2010; Verburg et al. 2009). These factors result in the need for monitoring strategies different from those developed for more commonly monitored LULC transitions such as large-scale deforestation and urban expansion (Ellis 2004). Here we present general considerations and a suite of methods for estimating net changes in GHG emissions that specifically address the conditions of smallholder agriculture. In the process we illustrate the relative trade-offs between costs of analysis, precision, and accuracy.

There are four basic steps required to calculate GHG emissions/removals from LULC change:

- *Determine change in LULC*. Changes in the areal extent of LULC classes must be determined by comparing data collected from two or more points in time.
- *Develop a baseline*. Observed changes in carbon stocks must be compared against a "business as usual" scenario of what would have happened in the absence of project activities. This step is generally carried out by either developing a baseline scenario or through direct observation of a reference region.
- *Calculate carbon stock changes*. Carbon stocks associated with LULC classes must be quantified for each point in time or emission factors must be used to calculate carbon stock changes and associated GHG emissions or removals.
- *Assess accuracy and calculate uncertainty*. Accuracy of each step must be assessed in order to determine the uncertainty associated with final emission/removal estimates associated with LULC changes.

It is important to note that these steps are not necessarily chronological. For example a baseline scenario could be developed prior to LULC change detection. Accuracy assessments should be done concurrently with each phase of data collection and analysis.

In order to carry out the above steps, two basic types of data are required, defined by the Intergovernmental Panel on Climate Change (IPCC) as activity data and emission factors (IPCC 2006). Activity data refer to the areal extent of chosen LULC categories, subcategories, and strata and are generally presented in hectares. Emission factors refer to the data used to calculate carbon stocks associated with activity data and are usually presented as metric tons of carbon (or carbon dioxide equivalents) per hectare. Emission factors may not be required for all carbon pools when carbon stock densities are inventoried directly using field sampling and/or remote sensing techniques. The IPCC Guidelines (2006) also lay out three tiers of methods used to calculate GHG emissions and reductions, which increase not only in precision and accuracy but also in data requirements and complexity of analysis. Tier 1 requires country-specific activity data but uses IPCC default emission factors that can be found in the IPCC Emission Factor Database (IPCC n.d.) and analysis is generally simple and of low cost. Tier 2 uses similar methods to Tier 1 but requires the use of some region- or country-specific emission factors or carbon stock data for key carbon pools and LULC categories (more information on key pools can be found in Sect. 3.4.1). Tier 3 requires high-resolution activity data combined with highly disaggregated inventory data for carbon stocks collected at the national or local level and repeated over time.

Collection of data to generate emission factors or calculate carbon stock densities is covered elsewhere in this book. The focus of this chapter is on the generation of activity data and the various methods available to link emission factors and/or carbon stock densities with activity data for estimating total carbon stocks and GHG emissions/removals at the landscape-scale. The following sections provide an overview of the general activities for each of the four steps required to calculate GHG emissions/reductions from LULC change, with a focus on smallholder agriculture landscapes. Trade-offs between uncertainty and cost are addressed and a variety of references—including existing protocols, scientific research, and review papers—are cited. Summary tables are presented at the beginning of each section, with a complete table at the end of the chapter (Table 3.8).

3.2 Determining Change in LULC

The IPCC Guidelines (2006) outline three specific Approaches to monitoring activity data (described in detail below). The three Approaches refer to the representation of land area and will influence the ability to meet the three IPCC Tiers, which indicate the overall uncertainty of GHG emission/reduction estimates (Table 3.1). In general, progressing from Approach 1 to 3 increases the amount of information associated with activity data but requires greater resources. It should be noted that increasing the information contained within activity data does not guarantee a reduction in uncertainty. Accuracy will ultimately depend on the quality of data and implementation of the Approach as much as the Approach itself (IPCC 2006). However, progressing from Approach 1 to 3 provides the opportunity for reducing uncertainty and meeting higher Tier requirements.

Approach 1 uses data on total land-use area for each LULC class and stratum but *without* data on conversions between land uses. The result of Approach 1 is usually a table of land-use areas at specific points in time and data often come from aggregated household surveys or census data. Results are not spatially explicit, only allow for the calculation of net area changes and do not allow for analysis of GHG emissions/removals for land remaining within a LULC category or the exploration of

Table 3.1 Summary of activities to determine change in LULC at various uncertainty levels

Activity	Higher uncertainty	Mid-range uncertainty	Lower uncertainty	Key references
Data acquisition	Approach 1 or 2 with minimal or no data collection (using existing aggregated datasets such as census or existing maps)	Approach 2 with disaggregated datasets (existing or developed)	Approach 3 with mid-resolution imagery and supplementary data	De Sy et al. (2012); IPCC (2006); Ravindranath and Ostwald (2008)
		Approach 3 with coarse or mid-resolution imagery	Approach 3 with very high-resolution imagery	
LULC classification	Broad LULC categories developed through subjective (non-empirical) survey methods; not spatially explicit	Broad LULC categories with simple subclasses or strata	Empirically derived LULC categories and strata	GOFC-GOLD (2014); IPCC (2006); Vinciková et al. (2010)
		Classified using visual interpretation or pixel-based techniques with limited or imagery-based training data; spatially explicit	Supervised classification using pixel-based, object-based or machine learning techniques with field-derived training data; spatially explicit	
LULC change detection	Arithmetic calculation of change in total land area for each LULC class using data generated by Approach 1	Arithmetic calculation of change in total land area for each LULC class and transitions between LULC classes using data generated by Approach 2 or; post-classification comparison with coarse or mid-resolution imagery	Spatially explicit change detection using post-classification comparison, image comparison, bitemporal classification or other GIS-based approaches	Huang and Song (2012); van Oort (2007)

drivers of LULC change. Therefore Approach 1 may not be suitable for carbon crediting under mechanisms such as the Verified Carbon Standard (VCS) or Reducing Emissions from Deforestation and Forest Degradation (REDD+) (see GOFC-GOLD 2014).

Approach 2 builds on Approach 1 by including information on conversions from one LULC class to another, but the data remain spatially non-explicit. This provides the ability to assess changes both into and out of a given LULC class and track conversions between LULC classes. A key benefit of Approach 2 is that emission factors can be modified (if data are available) to reflect specific conversions from one LULC category to another. For example, forests with a long history of prior cultivation may store less carbon than undisturbed forests of the same age (e.g., Eaton and Lawrence 2009; Houghton et al. 2012). Such factors cannot be taken into account using Approach 1. The results of Approach 2 can be expressed as a land-use conversion matrix of the areal extent of initial and final LULC categories.

Approach 3 uses datasets that are spatially explicit and compiled through sampling and wall-to-wall mapping techniques. Remotely sensed data (e.g., imagery from aerial- or satellite-based sensors) are often used in combination with georeferenced sampling such as field or household surveys. Data are then analyzed using geographic information systems (GIS) and can be easily combined with other spatially explicit datasets to stratify LULC categories and emission factors. This can greatly improve the accuracy of emission/removal estimates, especially for large areas, and allows for statistical quantification of uncertainty. Approach 3 can be an efficient way to monitor large areas. However it may require greater human and financial resources, which could be cost-prohibitive for smaller projects, especially if the spatial resolution of freely available or low-cost imagery is too coarse to detect LULC changes. (See Sect. 3.2.2 for more information about remotely sensed data.)

3.2.1 Setting Project Boundaries

The extent, location, and objectives of monitoring will all influence the appropriate choice of methods for analyzing LULC change and associated GHG emissions and reductions. While activity data may or may not be spatially explicit, the extent (i.e., boundaries) of the area monitored must be explicitly and unambiguously defined and should remain the same for all reporting periods. Several factors should be considered when defining the extent of the monitoring area.

Baseline Development and Data Availability. The availability of existing data (e.g., historical and/or cloud-free satellite imagery, forest inventories, research studies, census data) can determine the area for which a justifiable baseline scenario can be developed and therefore the project extent may need to be adjusted accordingly (Sect. 3.3). In some cases, it might be useful to adhere to political divisions rather than geographic boundaries if socioeconomic data are available in political units that do not correspond with geographic boundaries such as a watershed or ecoregion. If a reference region is to be used, it is important to consider whether one of appropriate size and characteristics can be found to match the chosen inventory extent (Sect. 3.3.2). For example the reference region may need to be 2–20 times larger than the project area to meet some VCS methodologies (VCS Association 2010).

IPCC Tier Selection. The inventory area may need to be reduced in order to meet higher IPCC Tier levels. For example, if a spatially explicit inventory (Approach 3) meeting IPCC Tier 3 guidelines is desired, expensive high-resolution satellite imagery and intensive data collection may be required and resource constraints may lead to a smaller inventory area. Meeting a lower IPCC Tier requirement could allow for the use of freely available imagery and/or existing data that could enable monitoring of a larger area.

Stratification and Variability. Ideally, inventory data will be collected in such a way as to sufficiently capture the spatial variability of key stratification variables. Identification of such variables a priori may reveal that it is impractical or financially unfeasible to develop a sampling strategy that can sufficiently capture variation within the entire area and the extent of the monitoring area may need to be adjusted.

Policy Levers. It is important to consider which policy levers exist, at what scale they can be applied and which may be influenced by assessment results when determining monitoring boundaries. For example, if regulations affecting land-use are implemented solely along political boundaries, it may not make sense to draw project-monitoring extents around watershed boundaries that may encompass multiple political units with differing regulations or policy options.

3.2.2 Data Acquisition

Data to estimate areal LULC extents can be acquired through three general sources: existing datasets developed for other purposes, collection of new data through sampling and complete LULC inventories using remote sensing data (Table 3.1).

Existing Data

Existing datasets can come from national or international sources or from other projects or research activities. Data may be available in a variety of formats and collection dates, and at varying spatial and temporal scales and extents. Time should be taken to identify existing data sources in order to determine what data remain to be collected, at what temporal and spatial scales and to what degree project resources can accommodate these needs. Useful datasets can include historical LULC maps, climate data, biophysical data (e.g., soil or hydrological maps), census or household surveys and political boundaries or administrative units.

Ground-Based Field Sampling Methods

Ground-based methods are recommended when existing datasets are incomplete, out of date, or inaccurate and complete spatial coverage with remote sensing techniques is unfeasible or would not be accurate on its own (IPCC 2003, Sect. 2.4.2). Ground-based sampling can be expensive and time consuming and is generally more appropriate for smaller project areas or when used in a sampling framework over larger areas. Field sampling to help determine LULC areal extents can result in two types of geographic data: biophysical data and socioeconomic data. Biophysical data generally require objective physical measurement of various land attributes (e.g., parcel size, vegetative composition). Ideally these measurements are georeferenced using GPS in order to integrate them with remote sensing data and enable accurate follow-up measurements. Socioeconomic data can be collected using a variety of methods including interviews, surveys, census, questionnaires, and

Box 3.1 Random and Targeted Sampling Methods for Generating LULC Activity Data

Random Sampling

Random sampling is generally done using systematic or stratified sampling methods. Systematic sampling spatially distributes sampling locations in a random but orderly way, for example using a grid. Stratified sampling selects sample sites based on any number of environmental, geographic, or socioeconomic variables to achieve sampling rates in proportion to the distribution of the chosen variables across the inventory extent. Stratified sampling methods (e.g., optimum allocation) can improve the accuracy and reduce costs of monitoring efforts (Maniatis and Mollicone 2010) and tools exist to determine the number of sample plots needed (UNFCCC/CCNUCC 2009). Ideally sample sites for determination of LULC can be co-located with sites for measuring carbon stocks and GHG emissions, although this may not always be practical or feasible.

Targeted Sampling

Targeted sampling refers to the non-random selection of specific sample regions based on determined criteria. A common example of targeted sampling is the use of low-cost or free-imagery to identify "hotspots" of active LULC change such as deforestation (Achard et al. 2002; De Sy et al. 2012). These hotspots, or a randomly selected subset within, can then be selected as sample units for more in-depth monitoring using higher-resolution imagery and/or comprehensive fieldwork. These data can then be used to train LULC classification algorithms and assess the accuracy of results obtained using medium or coarse resolution imagery. Regardless of the method chosen, sampling should be statistically sound and allow for the quantification of uncertainty.

participatory rural appraisals (e.g., semistructured interviews, transect walks, and other flexible approaches involving local communities; see Ravindranath and Ostwald 2008 for more information). Socioeconomic data may or may not be georeferenced, depending on the application.

Both biophysical and socioeconomic data acquired using the methods mentioned above can give a reasonable estimate of the proportions of LULC categories within the inventory area provided sample locations are selected using statistically rigorous methods to maintain consistency and minimize bias. These proportions can then be multiplied by the total land area to generate activity data. Sample locations can be chosen using random or targeted (non-random) methods (Box 3.1). Random methods allow for quantification of uncertainties and are therefore generally preferred, but targeted methods may be useful for measuring carbon stocks related to a specific event (e.g., a fire) or calibration of modelling for a specific carbon pool (e.g., effects of decomposition on soil carbon) (Maniatis and Mollicone 2010).

Remote Sensing Data

Complete wall-to-wall LULC inventories are generally carried out using a combination of remote sensing data and field-based sampling. Remotely sensed data come from aerial photography, satellite sensors, and airborne or satellite-based RADAR or LiDAR. Optical sensors are the most commonly used in LULC classification as they provide spectral information in the visible and infrared bands at a range of resolutions and costs (Table 3.2). While fine (<5 m) or medium (10–60 m) resolution imagery are preferable for accurately monitoring LULC in landscapes dominated by smallholder agriculture, cost of acquisition and/or processing may be prohibitive for projects covering large areas. However, methods exist for nesting high-resolution sampling within coarser resolution wall-to-wall coverage to reduce uncertainty of LULC change analysis across large areas and lower costs (e.g., Achard et al. 2002; Jain et al. 2013).

Image processing techniques can be applied to the remotely sensed data to enhance particular land-cover types, or enable more accurate stratification and classification, such as the calculation of the Normalized Difference Vegetation Index (NDVI), developing textural variables (e.g., Castillo-Gonzalez 2009) or principle component analysis (PCA). Imagery can also be classified into land-cover classes enabling easier manipulation in a GIS. Spatial analysis of remotely sensed data combined with environmental and/or socioeconomic variables can also create additional datasets to further enhance classification and stratification. Designating ecological or anthropogenic biomes (Ellis and Ramankutty 2008), calculating market accessibility (Chomitz and Gray 1996; Southworth et al. 2004) and identifying landscape mosaics (Messerli et al. 2009) are examples of such user-generated datasets to improve analysis of LULC change and explore drivers of change in smallholder landscapes.

Spatial Considerations

The spatial scale(s) at which data collection and analysis will take place is a key factor to consider when developing a monitoring and analysis program. Changing the scale at which analysis takes place can result in significantly different results, even when using the same dataset. The “optimal” scale of measurement and prediction is project-specific and may even vary for different steps of analysis (Lesschen et al. 2005). Complementary analysis at multiple scales may further improve accuracy (Messerli et al. 2009). A number of factors related to spatial scale should be considered to maintain transparency, and improve accuracy and efficiency of analysis.

The finest-scale unit of data is called a minimum information unit or minimum mapping unit (MIU or MMU). This is often the size of a small contiguous group of pixels for remote sensing data or the household for census data, although data may only be available aggregated to an administrative unit such as a village or municipality. To qualify for carbon credits, for example under the REDD+ mechanism, MMUs of <1–6 ha are generally required (De Sy et al. 2012; GOFC-GOLD 2014). In land-

Table 3.2 Overview of existing remote sensing data sources (adapted from GOFC-GOLD 2014)

Sensor and resolution	Examples of current optical sensors	Minimum mapping unit	Repeat rate	Cost	Utility for monitoring
Coarse (250–1000+ m)	AVHRR (1978–)	~100 ha	0.5–1 day	Low or free	Consistent pantropical annual monitoring to identify large clearings, major fires, and locate "hotspots" for further analysis with mid-resolution
	SPOT-VGT (1998–)				
	Terra/AQUAMODIS (2000–)	10–20 ha			
	Envisat-MERIS (2004–2012)				
	VIIRS (2012–)				
Medium (10–60 m)	Landsat TM or ETM+, Terra-ASTER	0.5–5 ha	16–18 days (Landsat)	Landsat & CBERS are free; for others: <$0.001/km^2 for historical data $0.02/km^2 to $0.5/km^2 for recent data	Primary tool to map deforestation and estimate area change; historical time series often available for baseline development
	IRS AWiFs or LISS III CBERS HRCCD		26 days (SPOT)		
	DMC				
	SPOT HRV				
	ALOS AVNIR-2				
	RADAR				
Fine (<5 m)	RapidEye	<0.1 ha	1–5 days (IKONOS and Quickbird)	High to very high $2–30 /km^2	Validation of results from coarser resolution analysis, and training of algorithms; in-depth analysis of "hotspots" or smaller project areas
	IKONOS				
	QuickBird[a]				
	GeoEye				
	WorldView				
	Pleiades				
	Aerial photos				
	LiDAR				

[a]Satellite was decommissioned in early 2015 but archived data are available

scapes dominated by smallholder agriculture, individual LULC parcels are often 0.5 ha or smaller. When using remote sensing data, it is preferable to have MIUs (e.g., pixels) that are significantly smaller than the average farm size to avoid mixed pixels that encompass multiple LULC categories. However methods of remote sensing analysis, such as spectral unmixing (Quintano et al. 2012) and hierarchical training with very high-resolution imagery (e.g., Jain et al. 2013) have been developed to attempt to deal with the issue of mixed pixels in coarser resolution imagery.

It is important to consider the scale of all available data to avoid mismatches that could lead to data management problems or wasted resources. Depending on the analysis methods used, data may have to be resampled to the coarsest available dataset. For example, it may be unnecessary to acquire a 5 m digital elevation model for stratification if it will be combined with 30 m Landsat data.

Temporal Considerations

Several temporal boundaries should be fixed established during the development of a monitoring methodology.

Historical reference period. If developing a baseline scenario from a historical reference period, this period must be specifically defined and appropriate for scenario development.

Monitoring period: The period for which changes in GHG emissions and reductions from LULC change are to be monitored.

Timing of monitoring: The schedule for monitoring to take place. Care should be taken to acquire imagery and/or carry out field sampling as close to the same time of year as possible for each monitoring period as interannual variability in vegetative cover and phenology may vary significantly in some locations (Huang and Song 2012; Serneels et al. 2001). Changes in carbon stocks from LULC change, such as declines in soil organic carbon (SOC) or vegetative regrowth, may not be linear within a monitoring period or may level off to zero-change within the period, also requiring appropriately timed sampling or modelling.

Monitoring frequency: The frequency of monitoring activities (e.g., imagery acquisition, field-sampling, surveys). Management strategies within a LULC category, for example cropping intensity, can have significant impacts on carbon stocks (e.g., Schmook 2010). More frequently, strategically timed data collection (i.e., sampling and/or image acquisition) is often required to detect changes in management strategies within an LULC category (De Sy et al. 2012; Jain et al. 2013; Smith et al. 2012). In most cases, particularly when dealing with remote sensing, increasing the temporal resolution of data (i.e., more frequent acquisition) necessitates declining spatial coverage and resolution (due to either technological or cost-prohibitive factors) and this trade-off must be considered when choosing between data sources.

LULC change definitions. The time period after which a change in LULC is considered permanent must be determined. For example, shifting cultivation, common practice in smallholder agriculture, results in cycles of cultivation and fallow periods that vary year to year, yet can resemble managed or secondary forest-cover when observed over the long term (Houghton et al. 2012). These temporary changes in land-cover (e.g., from annual cropping to secondary forests) can be misinterpreted as afforestation or deforestation depending on the timing of sampling or image acquisition if they are not considered across their entire cycle with sufficiently frequent measurements (DeFries et al. 2007). One approach to account for fluctuating carbon stocks associated with shifting cultivation is to calculate time-averaged carbon stocks for a given land-use system (Bruun et al. 2009; Palm et al. 2005).

Other considerations. Many studies have found that land-use is often influenced by land features. For example, farmers may choose to cultivate areas with fertile, carbon-rich soils (e.g., Aumtong et al. 2009; Ellis and Ramankutty 2008; Jiao et al. 2010) or reduce fallow periods when the soil fertility is high (Roder et al. 1995) and leave forests intact only in areas with poor soils. This preferential selection can make it difficult to determine that land-use is in fact causing a change in soil carbon stocks, and not the other way around (soil carbon stocks influencing land-use). Repeated sampling may be required to observe carbon stock changes resulting directly from LULC conversion (Bruun et al. 2009). The effects of prior land-use on future carbon sequestration potential may also be significant (see Eaton and Lawrence 2009; Hughes et al. 1999). While difficult to quantify, these delayed fluxes can be included when considering LULC transitions (e.g., a forest converted from agriculture may not store the same amount of carbon as a forest converted from a pasture). Finally, complications can arise from temporal mismatching, for example if biophysical or social data are collected in a separate time period from satellite imagery. There may be benefits from matching the timing of data acquisition on various factors (Rindfuss et al. 2004).

3.2.3 LULC Classification and Change Detection

LULC Category Definition

Regardless of the Approach used to generate activity data, LULC categories must be clearly and objectively established and LULC categories, subcategories, and strata should be mutually exclusive and totally exhaustive (Congalton 1991) with clear definitions of transitions from one class to another. (Note that sophisticated analysis methods using non-discrete, probabilistic or "fuzzy" classification do exist (e.g., Foody 1996; Southworth et al. 2004), but are beyond the scope of this chapter). For example, forests are generally defined based on a threshold value of minimum area, height and tree crown cover and the Designated National Authority (DNA) for each country can aid in defining LULC category definitions (GOFC-GOLD 2014). Objective definitions are especially important in smallholder landscapes where shifting cultivation and fallow rotations are common and transitions between LULC classes may not be straightforward. Furthermore, since smallholder landscapes often consist of small and heterogeneous land uses, it is possible that sampling points may fall into more than one LULC category. Systematic, transparent, and objective methods are needed to determine to which LULC category a sampling point belongs (Maniatis and Mollicone 2010).

The IPCC Agriculture, Forestry, and Other Land-Use (AFOLU) Guidelines (2006) define the following six broad land-use categories:

- Forest Land
- Cropland
- Grassland
- Wetlands
- Settlements
- Other Land

These top-level classes were designed to be broad enough to encompass all land areas in a country and allow for consistent and comparable reporting between countries. Monitoring activities can further divide these classes into conversion categories (i.e., Forest Land converted into Cropland, Wetlands converted into Settlements). For REDD+ GHG inventories and Tiers 2 and 3 reporting, it is likely that these top-level classes must be further divided into subcategories and/or stratified to allow for disaggregation of carbon stocks and improved estimation accuracy. Subcategories refer to unique LULCs within a category (e.g., secondary forest, within Forest Land) that impact emissions and for which data are available. Identification of subcategories can greatly reduce uncertainty of carbon stock estimates. For example, Asner et al. (2010) found that secondary forests held on average 60–70 % less carbon than intact forests in the Peruvian Amazon, and other studies have found similarly large differences in carbon stocks between forest types (e.g., Eaton and Lawrence 2009; Saatchi et al. 2007), highlighting the importance of forest subclasses. Secondary forests, a significant LULC class in smallholder landscapes, are estimated to make up more than half of tropical forested areas and can be an important source or sink of carbon (Eaton and Lawrence 2009; Houghton et al. 2012). Therefore, distinguishing between secondary forests, bush-fallows, and undisturbed forests, while often challenging, will likely result in more accurate carbon stock estimates.

Stratification within LULC categories and subcategories can be based on any number of factors significant to emission estimation such as climate, ecological zone, elevation, soil type, and census data (e.g., population, management practices) (see Stratification, below). Final LULC categories and strata will depend on project location, climate and ecological factors, data availability, analysis capacity, and other factors. Ideally, however, subcategories or strata can be aggregated to correspond with the six broad land-use categories listed above to maintain consistency between country or project inventories. Designation of LULC classes and strata will also depend on the IPCC Approach chosen to represent land-use area data. To meet Approaches 2 and 3, data on conversion between LULC categories and strata must be available, potentially limiting the number of possible subcategories and strata.

LULC Classification, Mapping, and Tabulation

Non-spatially explicit methods for collecting activity data (Approaches 1 and 2) result in tables of land area totals by LULC category for a given point in time. Depending on how data are collected, these results can be aggregated to political or geographic boundaries and incorporated into existing maps. The data themselves are not spatially explicit in their disaggregated form and therefore exact patterns of land-use cannot be interpreted within the spatial unit of aggregation (Table 3.1). The original data will generally come from LULC surveys, census data, existing maps or a combination of these. Therefore uncertainty associated with Approaches 1 and 2 will depend in large part on the quality of the sampling methods used to collect the original data. Costs could range greatly depending on the size of the project area, availability of existing data, heterogeneity of the landscape, and accessibility, but in general Approaches 1 and 2 can be low-cost options, especially for smaller projects.

Spatially explicit methods for generating activity data (Approach 3) use a combination of remote sensing and field-based sampling to develop a wall-to-wall classified LULC map with which LULC category areas can be totalled. Wall-to-wall maps provide the opportunity for interpolation between data points using GIS software and the development of spatially explicit polygons and/or individual pixels assigned to various LULC categories. In this manner activity data can be efficiently calculated, overlaid with ancillary data for stratification, and integrated with emission factors to quantify and analyze GHG emissions/reductions, their spatial variability, and drivers. Many methods exist to classify LULC, but they can be grouped into three main categories: visual interpretation, unsupervised classification, and supervised classification (Box 3.2). Additionally, a number of pre- and/or post-processing steps may also be required to ensure accurate results. Choice of classification methods and image processing will depend on available resources, technical expertise, imagery, location, and available software. Greater detail on specific methodologies is presented on the associated website. Whichever methods are chosen for preprocessing, classification, and post-processing, they should be transparent, repeatable by different analysts, and results should be assessed for accuracy (GOFC-GOLD 2014).

Stratification

Once LULC classes have been identified and imagery classified, stratification by one or more variables may be desirable to improve estimation of carbon stocks, GHG emissions and reductions, and/or baseline development. The primary goal of stratification is to minimize the variability of carbon stock estimates within LULC categories (Maniatis and Mollicone 2010). The most basic form of carbon stock stratification is the development of subcategories (e.g., secondary forest versus mature forest; tree crops versus annual crops). Additional datasets and/or more intensive sampling may be required to identify subcategories, which may increase costs,

Box 3.2 General LULC Classification Methods Using Remote Sensing Data

Visual interpretation
The simplest method of LULC classification is visual interpretation. In this method, a person familiar with the landscape and the appearance of LULC classes in remotely sensed imagery, manually interprets and classifies polygons around different land-covers. This method can be quite accurate but may not be precisely repeatable and can result in high uncertainty if comparisons are made between maps classified by different people. However systematic approaches to visual interpretation can increase accuracy and repeatability (e.g., Achard et al. 2002; Ellis 2004; Ellis et al. 2000).

Unsupervised classification
This method is fully automated and classification occurs without direct user intervention, although parameters such as the number of classes to be identified can be set by the user. Unsupervised classification algorithms cluster pixels into spectrally similar classes and very small spectral differences between classes can be identified (Vinciková et al. 2010). This method can be useful for exploring the number and distinguishability of potentially identifiable classes.

Supervised classification
Supervised classification relies on the training data that is used to calibrate automated or semiautomated classification algorithms. Training data may be obtained through field sampling, separate higher-resolution remote sensing imagery or from within the original image. Ideally training points will be chosen in a statistically rigorous way (e.g., random, stratified, systematic) and spatial and temporal factors should be considered (Sect. 3.2.2, Spatial Considerations and Temporal Considerations).

- *Pixel-based supervised classification.* Pixel-based supervised classification is one of the most commonly used classification methods. It uses spectral information for placing individual pixels into classes. Algorithms use training data and predetermined classes identified by the user to classify pixels. Statistical methods such as signature separability functions can be used to evaluate the quality of training of data and improve classification accuracy (Moreno and De Larriva 2012). One drawback to pixel-based classification, be it supervised or unsupervised, in smallholder agriculture landscapes is the problem of mixed pixels where individual pixels encompass multiple LULCs. Spectral mixture analysis (SMA), also called spectral unmixing, can overcome this problem by assigning individual pixels an estimated proportional value of multiple LULC classes (Quintano et al. 2012). SMA can improve classification accuracy in heterogeneous landscapes but requires significant technical expertise and expensive GIS software.
- *Object-based classification.* The primary goal of object-based classification is to identify MIUs on which to base classification criteria (Castillejo-

Box 3.2 (continued)

González et al. 2009). In pixel-based classification, the pixel is the MIU whereas object-based methods quantitatively group pixels that are spectrally similar and spatially adjacent to create new MIUs representing patches or parcels of homogenous land-covers. Classification is then carried out on individual objects using a combination of spatial and spectral information. Object-based techniques combined with high-resolution imagery have not only been shown to outperform pixel-based methods in highly heterogeneous landscapes (e.g., Moreno and De Larriva 2012; Perea et al. 2009) but also require extensive technical expertise, time, and specialized GIS software.

- *Other supervised classification techniques*—Additional, relatively complex techniques such as regression/decision trees, neural networks, hierarchical temporal memory (HTM) networks (Moreno and De Larriva 2012), and support vector machines (Huang and Song 2012) have also shown success in improving classification accuracy in heterogeneous landscapes.

and transparent objective methods should still be used to define subcategories. However, stratification can reduce overall costs if monitoring activities can be targeted toward subcategories in which LULC transitions and carbon stock changes are expected (GOFC-GOLD 2014). Further stratification can be done using biophysical (e.g., slope, rainfall, soil type) and socioeconomic (e.g., population) datasets. Combining datasets requires either spatially explicit data (Approach 3) or datasets following Approaches 1 or 2 that have been aggregated to spatially defined units such as administrative boundaries. (See Lesschen et al. (2005) for a good overview on combining datasets for analysis of LULC change in farming systems.)

Stratification should only be carried out to the degree that chosen strata improve carbon stock estimates and reduce uncertainty. Statistical methods such as multivariate and sensitivity analyses exist to assess the quality of potential strata. Project objectives, timeframe, and the temporal and spatial resolution of available data will also impact the choice of LULC subcategories and strata.

LULC Change Detection

When using activity data generated with Approaches 1 and 2, change detection can be as simple as carrying out basic arithmetic to calculate the change in total land area of each LULC class at two or more points in time. Approach 2 will include results on the specific transitions observed (e.g., from forest to cropland versus from forest to pasture) and results are generally reported using a land-use conversion matrix (IPCC 2006; Ravindranath and Ostwald 2008).

Spatially explicit methods (Approach 3) to detect changes in LULC can be separated into three general categories: post-classification comparison, image comparison approach, and bitemporal classification approach. Post-classification comparison is the most straightforward approach and consists of first conducting separate LULC classifications on two or more images and comparing the results to detect change. Post-classification change detection is popular due to the fact that hard classification for single-date imagery is often required for other purposes or preexisting classified images are being used for one or more dates (van Oort 2007). One major drawback to this approach is that each image will contain uncertainty stemming from misclassification, which could result in significant errors in the change map from misidentification of LULC change. The image comparison approach attempts to reduce these errors by comparing the two unclassified images and identifying pixel-based change thresholds through methods such as differencing, ratioing, regression, change vector analysis, and principal component analysis (Huang and Song 2012). Bitemporal classification goes a step further by analyzing multiple images simultaneously and applying one of a variety of algorithms to produce a final map with change classes in a one-step process (Huang and Song 2012). The two latter approaches can be more adept at detecting specific changes of interest and more subtle changes (van Oort 2007) and may reduce uncertainty in cases where classification accuracy is low.

3.3 Developing a Baseline

Activity data are monitored at two or more points in time to assess LULC change. However, this change must be compared against a "business as usual" scenario to determine additionality (i.e., to define what would have occurred in the absence of project interventions). Only by comparing observed changes against a well-developed and justified baseline can we be sure that project activities resulted in changes that would not have occurred otherwise. Two general methods exist to develop a comparative baseline of LULC change: the development of a baseline scenario or the monitoring of a reference region.

3.3.1 Baseline Scenarios

A baseline scenario predicts the LULC changes that would occur within the inventory area in the absence of interventions by creating a "business as usual" scenario from a variety of input data (Table 3.3). This scenario can be developed on a project-by-project basis using conditions and information particular to the project (project-specific approach) or for a specific geographic area, which may extend beyond the project area boundaries (regional baseline approach, also called the performance standard approach). Either approach can be based on historical data and/or logical arguments about economic opportunities that could influence future LULC change (Sathaye and Andrasko 2007) and examples of both approaches are given in

Table 3.3 Summary of activities for developing a baseline at various uncertainty levels

Activity	Higher uncertainty	Mid-range uncertainty	Lower uncertainty	Key references
Baseline scenario development	Logical arguments or simple trend analysis based on limited historical data	Projection of historical LULC trends using multitemporal historical data and/or simple predictor variables; or monitoring of a similar reference region	Modelled baseline developed using empirically derived predictor variables from multitemporal historical datasets; or	Brown et al. (2007); Greenhalgh et al. (2006); Sathaye and Andrasko (2007)
			Monitoring of a highly similar reference region with clearly defined comparative thresholds	
Baseline justification	Logical arguments and/or qualitative investment, barrier or common practice analysis	Investment, barrier and/or common practice analysis using limited quantitative analysis	Development of alternative baseline scenarios with investment and/or barrier analysis and common practice analysis using quantitative approaches	Greenhalgh et al. (2006); VCS Association (2012)

Table 3.4. The project-specific approach is often based on logical arguments where the baseline scenario is identified as the scenario facing the fewest barriers (Greenhalgh et al. 2006). This approach requires the development of multiple scenarios for the project area and requires economic-related data to evaluate which is most likely to occur. The regional baseline approach uses time-based estimates to project future carbon stock changes. This approach may require more GHG-related and spatially explicit data to enable quantitative analysis of trends in LULC change and GHG emissions/removals (Greenhalgh et al. 2006). The regional approach can result in more credible and transparent baselines and reduce costs when multiple projects are proposed within the same region (Brown et al. 2007; Sathaye and Andrasko 2007). An example of a potentially very useful dataset for identifying historical trends of forest-related disturbances is the high-resolution global forest change map recently published by Hansen et al. (2013).

Modelling future LULC changes based on historical and current data can be done using solely historical trends in percent change in land area or by incorporating drivers of LULC change into predictive models. Projection of historical LULC change trends requires reliable activity data for at least two points in time, preferably at the beginning and end of the historical period. Drivers used in modelled baselines can be simple metrics (e.g., population growth) to meet Tiers 1 and 2, or a more complex combination of spatially explicit biophysical and socioeconomic factors to meet Tiers 2 and 3. Drivers can greatly improve baseline development by capturing periodic fluctuations or variations across a landscape that may not be captured using trend analysis (Sathaye and Andrasko 2007). For example historical deforestation

Table 3.4 Overview of methods for baseline development

Method name	Scale of applicability	Method and data requirements	Advantages	Disadvantages	IPCC Tier	Key References
Alternative scenario development with barrier and common practice analysis	Project	Identify and evaluate potential baseline candidates based on biophysical conditions, socioeconomic and cultural factors and physical infrastructure	Adaptable to a variety of project activities, scales and resource constraints	Project-specific	Tier 1, 2 or 3	Greenhalgh et al. (2006); UNFCCC/ CCNUCC (2007); VCS Association (2012)
		Identify barriers to the project activity and baseline candidates		May not be spatially explicit		
Simple historical extrapolation	Project, region, country	Simple extrapolation of historical land-use change trends	Simple	Does not account for drivers of historical trends and assumes trends will continue	Tier 1 or 2	Lasco et al. (2007)
			Potentially low-cost	Not spatially explicit—applies the same trend to the entire area		
			Objective			
Forest Area Change (FAC)	Region, country	Ratio of non-forest area to total area	Minimal data requirements	Lack of spatial resolution	Tier 1 or 2	Brown et al. (2007)
		Population density		Reliance on only two major variables		
			Potentially low-cost; applicable to large regions	Limited applicability at fine spatial scales depending on data availability; applicable primarily to deforestation only		

Land-use carbon sequestration (LUCS)	Project, region, country	Rate of population growth and expected stabilization; initial area of principal land-uses; required agricultural land for population	Applicable at many scales; not restricted to deforestation	Lack of spatial resolution; complex model code and structure; assumptions needed for often poorly known parameters	Tier 1 or 2	Brown et al. (2007)
Geographical modelling (GEOMOD)	Project, region, country	Numerous spatial data layers of biophysical and socioeconomic factors	Spatially explicit results can be scaled to any resolution for which data are available	Large data requirements	Tier 3	Brown et al. (2007)
			Kappa for-location statistic can be calculated to evaluate model performance	Potentially high-cost		
				Must experiment with large number of variables to identify those providing the most explanatory power		

trends may not continue into the future if certain thresholds have been reached or land-use determinants such as road networks have changed (Chomitz and Gray 1996). Incorporating such factors into models can improve trend prediction and many different models exist to analyze the influence of drivers and set baselines (e.g., Brown et al. 2007). Reporting should describe the model and drivers in detail and the chosen model should be transparent, include empirical calibration and validation processes and generate uncertainty estimates (Greenhalgh et al. 2006).

To qualify for carbon crediting under the VCS, Clean Development Mechanism (CDM), REDD+ or other mechanisms, the baseline must generally be justified using investment, barrier and/or common practice analysis (Greenhalgh et al. 2006;Tomich et al. 2001; VCS Association 2012). In other words, barriers to the LULC changes sought by project activities or policies must be identified to show that insufficient incentives exist to achieve the desired LULC changes without intervention. Ideally multiple scenarios will be developed and evaluated to determine which is the most credible and conservative baseline choice. Several temporal considerations also exist related to both the historical period used to generate a baseline scenario and the period for which the baseline is projected forward. Historical data should be as relevant as possible to the projected period and major events (e.g., hurricanes, fires) and policy changes (e.g., protected area designations) should be considered when acquiring historical data. A narrative approach exploring the story behind historical LULC dynamics can further reveal relationships between observed changes and the forces driving them (Lambin et al. 2003). The validity period for the baseline (i.e., for how many years the baseline is considered valid and accurate) should also be taken into account. Experience from other projects suggests that an adjustable baseline approach is preferable. A common approach is to set a fixed baseline for the first 10 years, at which point it is evaluated and adjusted as needed (Brown et al. 2007; Sathaye and Andrasko 2007; VCS Association 2014).

3.3.2 *Reference Regions*

An alternative to developing a baseline scenario for the project area is to monitor a separate reference region, a common approach among Voluntary Carbon Standard (VCS) methodologies (e.g., VCS Association 2010 and others). The reference region should be sufficiently similar to the project area to conclude that the trajectory of LULC change observed in the reference region would also have occurred within the project area in the absence of project activities. While exact requirements for identification of a reference region vary, in general the reference region must be significantly larger than and demonstrably similar to the inventory area. In order to demonstrate similarity, key variables must be compared which may include landscape features (e.g., slope, elevation, LULC distribution), ecological variables (e.g., rainfall, temperature, soil type) and socioeconomic conditions (e.g., population, land tenure status, policies, and regulations) (see VCS Association 2010). Transparent comparison procedures must be developed to set comparative thresholds for the reference region (e.g., average slope of the reference region shall be within 10 % of the average slope of the inventory area).

Monitoring a reference region may be a cost-effective option for small projects that can easily identify an area similar to the project area. However larger projects, or projects working in a unique biophysical or sociopolitical environment, may find it difficult to locate an appropriate reference region, or may find it cost-prohibitive to monitor one.

3.4 Calculating Carbon Stock Changes

In order to estimate GHG emissions and removals, carbon stock densities must be quantified for each LULC category subclass and/or stratum. Carbon stock densities may come from default values, national datasets, scientific studies or field sampling and are generally given as tons of carbon per hectare (Mg C ha^{-1}) for individual or combined carbon pools (Table 3.5).

3.4.1 Key Carbon Pools

The IPCC Guidelines (2006) define five carbon pools: living aboveground biomass, living belowground biomass, deadwood, litter and soil organic matter (SOM). In the case that data are not available for all carbon pools, key pools can be identified based on their relative expected contribution to total carbon stock changes caused by possible LULC transitions. Thresholds are developed to delimit the minimum contribution of total emissions from a pool to be defined as "key." For example, a threshold could be created stating that only pools representing more than 10 % of total carbon stocks are considered key. Therefore it is possible that some pools will be key for certain LULC classes but not for others. Identifying key pools can help target monitoring and modelling efforts to minimize uncertainty and is required under IPCC reporting.

3.4.2 Initial Carbon Stock Estimates

Calculation of initial carbon stocks can be done in several ways ranging from the use of simple arithmetic to running complex models. The simplest method is to assign a single carbon stock density value (or range of values) to each LULC category and multiply this value by the total area of each category. This method can be used with activity data associated with any of the three Approaches. It is relatively straightforward and potentially low-cost, but may introduce high levels of uncertainty as it assumes that there is no variability of carbon stocks within LULC categories.

Uncertainty can be reduced by taking into account additional drivers of carbon stocks beyond just LULC categories. This can be done through stratification (Sect. 3.2.3) and/or modelling. Modelling approaches require data on carbon stocks and rates of change, which can be obtained from default emission factors, scientific research, or field measurements. Additional biophysical (e.g., slope, rainfall, soil type) and socioeconomic (e.g., population) datasets may also be needed. A variety

Table 3.5 Summary of activities for calculating carbon stock changes from LULC change at various uncertainty levels

Activity	Higher uncertainty	Mid-range uncertainty	Lower uncertainty	Key references
Define key carbon pools	Key pools identified using international or default data;	Key pools identified using region-specific or field-based data	Key pools identified for each LULC class using field sampling, or	GOFC-GOLD (2014); IPCC (2006, Volume 4, Chap. 2)
	Same key pools applied to all LULC classes	Key pools defined separately for at least broad LULC categories	Data available for all carbon pools	
Initial carbon stock estimates	Single carbon stock density applied to each LULC class based on global or regional default data	Carbon stocks stratified by subclasses or additional strata and derived from country-specific data and/or field sampling for key carbon pools	Spatially explicit stratification and modelling of carbon stocks using empirically derived drivers of observed carbon stock variability; or	Goetz et al. (2009); GOFC-GOLD (2014); Greenhalgh et al. (2006); IPCC (2006)
			Direct carbon stock monitoring approaches (e.g., using LiDAR, RADAR, optical sensors)	
Monitoring carbon stock changes	Process-based method using default emissions factors assigned to LULC classes and change processes (e.g., deforestation)	Process-based method using emission factors derived from country- or region-specific data	Process-based method using emission factors derived from field sampling within the project area or research activities in highly similar areas	Greenhalgh et al. (2006); Houghton et al. (2012); IPCC (2006, Volume 4, Chap. 2)
			Stock-based methods using multitemporal carbon stock inventories for key pools	

of models such as PROCOMAP, CO_2FIX, CENTURY, ROTH, and others exist with a range of complexity and data requirements. (See Ravindranath and Ostwald 2008 for a good comparison of several models.)

3.4.3 *Monitoring Carbon Stock Changes*

Carbon stock changes are estimated using one of two general methods: one process-based and the other stock-based. The process-based method estimates the net additions to, or removals from, each carbon pool based on processes and activities that result in carbon stock changes, such as tree harvesting, fires, etc. The stock-based method estimates emissions and removals by measuring carbon stocks in key pools at two or more points in time.

Process-Based Method

The process-based method (sometimes called the gain-loss, IPCC default or emission factor method) estimates gains or losses of carbon in each pool by simulating changes resulting from disturbance or recovery (Houghton et al. 2012). Changes in LULC drive process-based models, and carbon stocks are re-allocated based on observed or modelled LULC change. Gains are a result of carbon accumulation from the atmosphere (e.g., in tree biomass) or transfers from another pool (e.g., from biomass to SOC via decomposition). Losses are attributed to transfers to another pool or emissions to the atmosphere as CO_2 or other GHGs (IPCC 2006, Volume 4, Chap. 2). Additional emission factors can be developed for emitting activities that do not necessarily affect the five carbon pools identified by the IPCC. These include, for example, direct emissions from livestock, farm equipment or the production of non-food products. Models and emission factors used in process-based methods can vary in complexity and potentially meet any Tier requirements. IPCC default factors can be used to achieve Tier 1 reporting requirements whereas country-specific or locally derived research data combined with more complex modelling approaches are required to meet Tier 2 and 3 requirements.

Stock-Based Method

The stock-based method (also called the bookkeeping, stock-difference, or stock-change method) combines ground-based and/or remotely sensed data of measured carbon stocks with data on changes in the total land area of each LULC class between two or more points in time. For stock-based methods, carbon stock changes are measured independently of LULC change and are then multiplied by the total area of each LULC class and stratum. Process-based methods model carbon stock changes based on LULC changes. Depending on the spatial resolution of data, conversions might be required to arrive at a carbon density (Mg C ha^{-1}) that is then combined with activity data to estimate total emissions/removals. Typically, country-specific information is required for use with the stock-based method and resource requirements for data collection may be greater than process-based methods unless appropriate datasets already exist. Stock-based methods often meet at least Tier 2 requirements, provided activity data were generated according to Approach 2 or 3.

3.5 Assessing Accuracy and Calculating Uncertainty

In order to qualify for carbon crediting under mechanisms such as VCS, CDM, and REDD+, final reporting of GHG emissions/removals associated with LULC change must include uncertainty estimates (Maniatis and Mollicone 2010). Uncertainty should be reported as the range within which the mean value lies for a given probability (e.g., a 95 % confidence interval) or the percent uncertainty of the mean value, each of which can be calculated from the other (IPCC 2003). Errors will be introduced at every level of data collection. Analysis and assessment of accuracy and uncertainty should be carried out for each step. Not only is this important for reporting purposes, it can provide valuable information to project managers to determine which steps contain the greatest sources of uncertainty, thereby encouraging cost-effective monitoring (e.g., Smits et al. 1999).

In this chapter we focus on estimating uncertainty associated with the collection of activity data, detection of LULC changes, and linking of emission factors and/or carbon stocks. Methods for assessing uncertainty related to the production of emission factors and measurement of carbon stocks (e.g., calculating soil carbon in a forest) are discussed elsewhere.

3.5.1 LULC Classification Accuracy Assessment

When remote sensing data are used to develop wall-to-wall LULC maps, two types of error exist: errors of inclusion (commission errors) and errors of exclusion (omission errors). Accuracy should be assessed using a statistically valid method, the most common method being statistical sampling of independent higher-quality validation sample units (e.g., pixels, polygons, sites) for comparison against classified sample units (Congalton 1991) (Table 3.6). These validation samples can be taken from field observations, additional higher-resolution remote sensing imagery, or can be visually identified from within the original image provided they are independent from those used during training. As with the selection of training data, validation sampling should be done in a statistically sound and transparent manner. Stratified or proportional sampling techniques may be desirable to improve accuracy and reduce costs. When using field-based sampling to analyze current imagery, validation data should be collected as close to the time of image acquisition as possible, ideally at the same time as training data. Including farmers or other community members in the data collection process can be an effective way to estimate past LULC for classification and validation of historical imagery, while at the same time empowering stakeholders and addressing conservation issues (e.g., Sydenstricker-Neto et al. 2004).

The accuracy of classified sample units compared against "real-world" validation sample units can be presented in an error matrix, also called a confusion matrix. This helps visualize errors, identify relationships between errors and LULC categories, and calculate indices of accuracy and variation (Congalton 1991). Classification accuracy refers to the percentage of sample units correctly classified and can be calculated as commission and omission errors for each LULC class as well as an overall accuracy for all classes (Table 3.7). These classification accuracies can then be used as an uncertainty estimate to discount carbon credits associated with LULC change. For example, to maintain conservativeness of carbon credit estimates the VCS Association VM0006 (2010) uses the smallest accuracy of all maps as a discount factor for carbon credits. In the hypothetical example from Table 3.7, this would result in carbon credits being discounted by 25 % (multiplied by a discount factor of 0.75). Representing accuracy using an error matrix also provides an opportunity to assess which LULC categories are most often confused. For example, cropland in smallholder landscapes

Table 3.6 Summary of activities for assessing accuracy and calculating uncertainty at various uncertainty levels

Activity	Higher uncertainty	Mid-range uncertainty	Lower uncertainty	Key references
LULC area estimates and change detection	Assessment of data collection procedures to ensure data quality, but without the use of methods to quantify uncertainty	Assessment of data quality through systematic analysis of data collection procedures; or error matrix with Kappa coefficient based on validation points from limited field ground-truthing or marginally higher-quality imagery	Confusion matrix with Kappa coefficient based on validation points from ground-truthing in the field or higher-quality imagery	Congalton (1991); IPCC (2006, Volume 4, Chap. 3)
			Calculation of confidence intervals for LULC category areas and changes in area	
Carbon stock estimates	Varies by carbon pool; See Chaps. 6 and 7 for more information			
Combining uncertainty estimates	Simple error propagation	Error propagation using more complex equations and controlling for correlation of input data	Monte Carlo simulations or other bootstrapping techniques	GOFC-GOLD (2014); IPCC (2003); Ravindranath and Ostwald (2008); Saatchi et al. (2007)

Table 3.7 Hypothetical error matrix showing the number of pixels mapped and validated (ground-truthed) by LULC class. Values in bold highlight the number of correctly mapped pixels and the row and column totals, which are used to calculate producer's and user's accuracy

Mapped classes	Ground truth classes						
	Forest	Cropland	Grassland	Wetland	Settlements	Other land	Total
Forest	**900**	50	50	0	0	0	**1000**
Cropland	50	**750**	150	30	20	0	**1000**
Grassland	30	60	**810**	70	20	10	**1000**
Wetland	30	30	30	**390**	0	20	**500**
Settlements	0	20	20	10	**420**	30	**500**
Other land	0	20	0	0	30	**450**	**500**
Total	**1010**	**930**	**1060**	**500**	**490**	**510**	**4500**

	Producer's accuracy (omission error)		User's accuracy (commission error)			
Forest	900/1010	89 %	900/1000	90 %		
Cropland	750/930	81 %	750/1000	75 %	Overall accuracy	
Grassland	810/1060	76 %	810/1000	81 %	3720/4500	83 %
Wetland	390/500	78 %	390/500	78 %		
Settlements	420/490	86 %	420/500	84 %		
Other land	450/510	88 %	450/500	90 %		

is often misclassified due to small farm sizes and its resemblance to bare soil (due to minimal reflectance from young crops) or secondary forests (due to intercropping with tree species commonly found in secondary forests) (e.g., Sydenstricker-Neto et al. 2004). Other accuracy indicators include the kappa coefficient or KHAT statistic, root mean squared error (RMSE), adjusted R^2, Spearman's rank coefficient and others (Congalton 1991; Jain et al. 2013; Lesschen et al. 2005; Smits et al. 1999).

3.5.2 LULC Change Detection Accuracy Assessment

The accuracy of LULC change detection can be assessed using methods similar to those used to validate single scene LULC classification, but additional considerations exist. When making post-classification comparisons using two independently classified images, the accuracy of each individual classification should be assessed in addition to the accuracy of the change image. It is usually easier to identify errors of commission in change products because often only a small proportion of the land area will have experienced change, and often within a limited geographic area (GOFC-GOLD 2014). Unique sampling methodologies may therefore prove more cost-effective to validate the relatively rare event of changes in LULC within an image (Lowell 2001). A transition error matrix can be used to report the accuracy with which transitions between LULC categories are detected. This allows for assessment of uncertainty for each transition (e.g., forest to cropland, forest to grassland) and for partitioning of uncertainty attributable to the change detection process versus classification (van Oort 2007).

3.5.3 Uncertainty Associated with Estimating Carbon Stocks

Uncertainty estimates should be developed for key carbon pools within each LULC category. Uncertainty of carbon stocks using the stock-based method will be related to sampling. The process-based method will contain uncertainty estimates derived from scientific literature, model accuracy or other sources. Factors such as the scale of aggregation, stratification variables, and the spatial or temporal considerations discussed above can all influence the uncertainty associated with integrating carbon stocks and activity data.

3.5.4 Combining Uncertainty Values and Reporting Total Uncertainty

Combining uncertainty estimates for activity data, LULC change detection and emissions factors or carbon stocks can be done several ways, ranging from simple error propagation calculations (Tier 1) to more complex Monte Carlo simulations, also called bootstrapping or bagging (Tiers 2 and 3). Several approaches exist for calculating error propagation. For example, different equations are recommended if input data are correlated (e.g., the same activity data or emission factors were used to calculate multiple input factors that are to be summed) or if individual uncertainty values are high (e.g., greater than 30 %) (GOFC-GOLD 2014; IPCC 2003). Monte Carlo simulations select random values within probability distribution functions (PDF) developed for activity data and associated carbon stock estimates to calculate corresponding changes in carbon stocks. The PDFs represent the variability of the input variables and the simulation is undertaken many times to produce a mean carbon stock-change value and range of uncertainty (see IPCC 2003 and citations within for more detailed information on running Monte Carlo simulations). Simulation results can be combined with classification accuracies to compute uncertainties for each pixel. This allows exploration of the variation of accuracy by LULC class or stratum, and where to target future measurements to achieve the greatest reductions in overall uncertainty (Saatchi et al. 2007). Generally speaking, Monte Carlo simulations require greater resources than error propagation equations, but both methods require quantitative uncertainty estimates for activity data, LULC changes, and carbon stocks.

3.6 Challenges, Limitations, and Emerging Technologies

Monitoring LULC change and associated GHG emissions/reductions in a cost-effective manner remains a challenge in heterogeneous landscapes such as those dominated by smallholder agriculture. Monitoring change in management within LULC categories can be even more challenging, yet management is often a key component of smallholder carbon projects. Technologies are emerging to directly monitor carbon stocks (namely aboveground biomass), which could overcome some of these challenges. For example LiDAR shows promise for accurate direct estimation of vegetation structure, aboveground biomass, and carbon stocks (Goetz and Dubayah 2011; Goetz et al. 2009). While direct measurement methods are generally still in the research phase and may be cost-prohibitive for most projects, they may prove especially useful for smallholder settings as they can improve accuracy by removing the error associated with misclassification of LULC, a potentially large source of uncertainty in heterogeneous landscapes. In the end, it is difficult to recommend a single methodological approach to monitoring LULC in smallholder landscapes as optimal methods will depend on the project area, size, available resources, time period, interventions, and other factors. An overall summary of the general methods discussed in each section of this chapter is presented in Table 3.8. Time should be taken to assess these methods and their associated trade-offs, read the relevant key references and stay abreast of emerging remote sensing options to identify the most appropriate methodology for specific project conditions.

Table 3.8 Overall summary of general methods to achieve various levels of uncertainty when estimating carbon stock changes resulting from LULC change

General method	Specific activity	Higher uncertainty	Mid-range uncertainty	Lower uncertainty	Key references
Determine change in LULC	Data acquisition	Approach 1 or 2 with minimal or no data collection (using existing aggregated datasets such as census or existing maps)	Approach 2 with disaggregated datasets (existing or developed)	Approach 3 with mid-resolution imagery and supplementary data	De Sy et al. (2012); IPCC (2006); Ravindranath and Ostwald (2008)
			Approach 3 with coarse or mid-resolution imagery	Approach 3 with very high-resolution imagery	
	LULC classification	Broad LULC categories developed through subjective (non-empirical) survey methods; not spatially explicit	Broad LULC categories with simple subclasses or strata	Empirically derived LULC categories and strata	GOFC-GOLD (2014); IPCC (2006); Vinciková et al. (2010)
			Classified using visual interpretation or pixel-based techniques with limited or imagery-based training data; spatially explicit	Supervised classification using pixel-based, object-based or machine learning techniques with field-derived training data; spatially explicit	
	LULC change detection	Arithmetic calculation of change in total land area for each LULC class using data generated by Approach 1	Arithmetic calculation of change in total land area for each LULC class and transitions between LULC classes using data generated by Approach 2 or; post-classification comparison with coarse or mid-resolution imagery	Spatially explicit change detection using post-classification comparison, image comparison, bitemporal classification, or other GIS-based approaches	Huang and Song (2012); van Oort (2007)

General method	Specific activity	Higher uncertainty	Mid-range uncertainty	Lower uncertainty	Key references
Develop a baseline	Baseline scenario development	Logical arguments or simple trend analysis based on limited historical data	Projection of historical LULC trends using multitemporal historical data and/or simple predictor variables; or monitoring of a similar reference region	Modelled baseline developed using empirically derived predictor variables from multitemporal historical datasets; or	Brown et al. (2007); Greenhalgh et al. (2006); Sathaye and Andrasko (2007)
				Monitoring of a highly similar reference region with clearly defined comparative thresholds	
	Baseline justification	Logical arguments and/or qualitative investment, barrier, or common practice analysis	Investment, barrier, and/or common practice analysis using limited quantitative analysis	Development of alternative baseline scenarios with investment and/or barrier analysis and common practice analysis using quantitative approaches	Greenhalgh et al. (2006); VCS Association (2012)
Calculate carbon stock changes	Define key carbon pools	Key pools identified using international or default data	Key pools identified using region-specific or field-based data	Key pools identified for each LULC class using field sampling, or	GOFC-GOLD (2014); IPCC (2006, Volume 4, Chap. 2)
		Same key pools applied to all LULC classes	Key pools defined separately for at least broad LULC categories	Data available for all carbon pools	
	Initial carbon stock estimates	Single carbon stock density applied to each LULC class based on global or regional default data	Carbon stocks stratified by subclasses or additional strata and derived from country-specific data and/or field sampling for key carbon pools	Spatially explicit stratification and modelling of carbon stocks using empirically derived drivers of observed carbon stock variability; or	Goetz et al. (2009); GOFC-GOLD (2014); Greenhalgh et al. (2006); IPCC (2006)
				Direct carbon stock monitoring approaches (e.g., using LiDAR, RADAR, optical sensors)	
	Monitoring carbon stock changes	Process-based method using default emissions factors assigned to LULC classes and change processes (e.g., deforestation)	Process-based method using emission factors derived from country- or region-specific data	Process-based method using emission factors derived from field sampling within the project area or research activities in highly similar areas	Greenhalgh et al. (20060; Houghton et al. (2012); IPCC (2006, Volume 4, Chap. 2)
				Stock-based methods using multitemporal carbon stock inventories for key pools	

(continued)

Table 3.8 (continued)

General method	Specific activity	Higher uncertainty	Mid-range uncertainty	Lower uncertainty	Key references
Assess accuracy and calculate uncertainty	LULC area estimates and change detection	Assessment of data collection procedures to ensure data quality, but without the use of methods to quantify uncertainty	Assessment of data quality through systematic analysis of data collection procedures; or error matrix with Kappa coefficient based on validation points from limited field ground-truthing or marginally higher-quality imagery	Confusion matrix with Kappa coefficient based on validation points from ground-truthing in the field or higher-quality imagery	Congalton (1991); IPCC (2006, Volume 4, Chap. 3)
				Calculation of confidence intervals for LULC category areas and changes in area	
	Carbon stock estimates	Varies by carbon pool; See Chaps. 6 and 7 for more information			
	Combining uncertainty estimates	Simple error propagation	Error propagation using more complex equations and controlling for correlation of input data	Monte Carlo simulations or other bootstrapping techniques	GOFC-GOLD (2014); IPCC (2003); Ravindranath and Ostwald (2008); Saatchi et al. (2007)

References

Achard F, Eva HD, Stibig H-J, Mayaux P, Gallego J, Richards T, Malingreau J-P (2002) Determination of deforestation rates of the world's humid tropical forests. Science 297:999–1002. doi:10.1126/science.1070656

Asner GP, Powell GVN, Mascaro J, Knapp DE, Clark JK, Jacobson J, Kennedy-Bowdoin T, Balaji A, Paez-Acosta G, Victoria E, Secada L, Valqui M, Hughes RF (2010) High-resolution forest carbon stocks and emissions in the Amazon. Proc Natl Acad Sci U S A 107:16738–16742. doi:10.1073/pnas.1004875107

Aumtong S, Magid J, Bruun S, de Neergaard A (2009) Relating soil carbon fractions to land use in sloping uplands in northern Thailand. Agric Ecosyst Environ 131:229–239. doi:10.1016/j.agee.2009.01.013

Brown S, Hall M, Andrasko K, Ruiz F, Marzoli W, Guerrero G, Masera O, Dushku A, DeJong B, Cornell J (2007) Baselines for land-use change in the tropics: application to avoided deforestation projects. Mitig Adapt Strateg Glob Chang 12:1001–1026. doi:10.1007/s11027-006-9062-5

Bruun TB, Neergaard A, Lawrence D, Ziegler AD (2009) Environmental consequences of the demise in swidden cultivation in Southeast Asia: carbon storage and soil quality. Hum Ecol 37:375–388. doi:10.1007/s10745-009-9257-y

Castillejo-González IL, López-Granados F, García-Ferrer A, Peña-Barragán JM, Jurado-Expósito M, de la Orden MS, González-Audicana M (2009) Object- and pixel-based analysis for mapping crops and their agro-environmental associated measures using QuickBird imagery. Comput Electron Agric 68:207–215. doi:10.1016/j.compag.2009.06.004

Chomitz KM, Gray D (1996) Roads, land use, and deforestation: a spatial model applied to Belize. World Bank Econ Rev 10:487–512. doi:10.1093/wber/10.3.487

Congalton RG (1991) A review of assessing the accuracy of classifications of remotely sensed data. Remote Sens Environ 37:35–46. doi:10.1016/0034-4257(91)90048-B

De Sy V, Herold M, Achard F, Asner GP, Held A, Kellndorfer J, Verbesselt J (2012) Synergies of multiple remote sensing data sources for REDD+ monitoring. Curr Opin Environ Sustain 4:696–706. doi:10.1016/j.cosust.2012.09.013

DeFries R, Achard F, Brown S, Herold M, Murdiyarso D, Schlamadinger B, de Souza C (2007) Earth observations for estimating greenhouse gas emissions from deforestation in developing countries. Environ Sci Pol 10:385–394. doi:10.1016/j.envsci.2007.01.010

Eaton JM, Lawrence D (2009) Loss of carbon sequestration potential after several decades of shifting cultivation in the Southern Yucatán. For Ecol Manage 258:949–958. doi:10.1016/j.foreco.2008.10.019

Ellis EC (2004) Long-term ecological changes in the densely populated rural landscapes of China. In: DeFries RS, Asner GP, Houghton RA (eds) Ecosystems and land use change. American Geophysical Union, Washington, DC, pp 303–320. doi:10.1029/153GM23

Ellis EC, Ramankutty N (2008) Putting people in the map: anthropogenic biomes of the world. Front Ecol Environ 6:439–447. doi:10.1890/070062

Ellis EC, Li RG, Yang LZ, Cheng X (2000) Long-term change in village-scale ecosystems in China using landscape and statistical methods. Ecol Appl 10:1057–1073. doi:10.2307/2641017

Foody GM (1996) Approaches for the production and evaluation of fuzzy land cover classifications from remotely-sensed data. Int J Remote Sens 17:1317–1340. doi:10.1080/01431169608948706

Goetz S, Dubayah R (2011) Advances in remote sensing technology and implications for measuring and monitoring forest carbon stocks and change. Carbon Manag 2:231–244. doi:10.4155/cmt.11.18

Goetz SJ, Baccini A, Laporte N, Johns T, Walker W, Kellndorfer J, Houghton R, Sun M (2009) Mapping and monitoring carbon stocks with satellite observations: a comparison of methods. Carbon Balance Manag 4:1–7. doi:10.1186/1750-0680-4-2

GOFC-GOLD (2014) A sourcebook of methods and procedures for monitoring and reporting anthropogenic greenhouse gas emissions and removals associated with deforestation, gains and losses of carbon stocks in forests remaining forests, and forestation. GOFC-GOLD Report version COP20-1, GOFC-GOLD Land Cover Project Office, Wageningen University, The Netherlands

Greenhalgh S, Daviet F, Weninger E (2006) The land use, land-use change, and forestry guidance for GHG project accounting. World Resources Institute, Washington, DC

Hansen MC, Potapov PV, Moore R, Hancher M, Turubanova SA, Tyukavina A, Thau D, Stehman SV, Goetz SJ, Loveland TR, Kommareddy A, Egorov A, Chini L, Justice CO, Townshend JRG (2013) High-resolution global maps of 21st-century forest cover change. Science 342(6160):850–853. doi:10.1126/science.1244693

Houghton R, House JI, Pongratz J, van der Werf GR, DeFries RS, Hansen MC, Le Quéré C, Ramankutty N (2012) Carbon emissions from land use and land-cover change. Biogeosciences 9:5125–5142. doi:10.5194/bg-9-5125-2012

Huang C, Song K (2012) Forest-cover change detection using support vector machines. In: Giri CP (ed) Remote sensing of land use and land cover, remote sensing applications series. CRC Press, Boca Raton, pp 191–206. doi:10.1201/b11964-16

Hughes R, Kauffman J, Jaramillo V (1999) Biomass, carbon, and nutrient dynamics of secondary forests in a humid tropical region of Mexico. Ecology 80:1892–1907

IPCC (2003) Good practice guidance for land use, land-use change and forestry. Institute for Global Environmental Strategies (IGES), Kanagawa, Japan

IPCC (2006) IPCC guidelines for national greenhouse gas inventories, prepared by the National Greenhouse Gas Inventories Programme. IGES, Geneva

IPCC (n.d.) Emissions Factor Data Base (EFDB). http://www.ipcc-nggip.iges.or.jp/EFDB/main.php. Accessed 14 March 2015

Jain M, Mondal P, DeFries RS, Small C, Galford GL (2013) Mapping cropping intensity of smallholder farms: a comparison of methods using multiple sensors. Remote Sens Environ 134:210–223. doi:10.1016/j.rse.2013.02.029

Jiao J-G, Yang L-Z, Wu J-X, Wang H-Q, Li H-X, Ellis EC (2010) Land use and soil organic carbon in China's village landscapes. Pedosphere 20:1–14. doi:10.1016/S1002-0160(09)60277-0

Lambin EF, Geist HJ, Lepers E (2003) Dynamics of land use and land cover change in tropical regions. Annu Rev Environ Resour 28:205–241. doi:10.1146/annurev.energy.28.050302.105459

Lasco RD, Pulhin FB, Sales RF (2007) Analysis of leakage in carbon sequestration projects in forestry: a case study of upper Magat watershed, Philippines. Mitig Adapt Strateg Glob Chang 12:1189–1211. doi:10.1007/s11027-006-9059-0

Lesschen JP, Verburg PH, Staal SJ (2005) Statistical methods for analysing the spatial dimension of changes in land use and farming systems. LUCC Report Series 7. International Geosphere–Biosphere Programme (IGBP), Nairobi

Lowell K (2001) An area-based accuracy assessment methodology for digital change maps. Int J Remote Sens 22:3571–3596. doi:10.1080/01431160010031270

Maniatis D, Mollicone D (2010) Options for sampling and stratification for national forest inventories to implement REDD+ under the UNFCCC. Carbon Balance Manag 5:1–14. doi:10.1186/1750-0680-5-9

Messerli P, Heinimann A, Epprecht M (2009) Finding homogeneity in heterogeneity—a new approach to quantifying landscape mosaics developed for the Lao PDR. Hum Ecol 37:291–304. doi:10.1007/s10745-009-9238-1

Moreno AJP, De Larriva JEM (2012) Comparison between new digital image classification methods and traditional methods for land-cover mapping. In: Giri CP (ed) Remote sensing of land use and land cover. CRC Press, Boca Raton, pp 137–152. doi:10.1201/b11964-13

Palm CA, van Noordwijk M, Woomer P, Alegre JC, Arévalo L, Castilla CE, Cordeiro DG, Hairiah K, Kotto-Same J, Moukam A, Parton WJ, Ricse A, Rodrigues V, Sitompul SM (2005) Carbon losses and sequestration with land use change in the humid tropics. In: Palm CA, Vosti SA, Sanchez PA, Ericksen PJ (eds) Slash-and-burn agriculture: the search for alternatives. Columbia University Press, New York, pp 41–63

Perea A, Meroño J, Aguilera M (2009) Algorithms of expert classification applied in Quickbird satellite images for land use mapping. Chilean J Agric Res 69:400–405

Quintano C, Fernández-Manso A, Shimabukuro YE, Pereira G (2012) Spectral unmixing. Int J Remote Sens 33:5307–5340

Ravindranath N, Ostwald M (2008) Carbon inventory methods: handbook for greenhouse gas inventory, carbon mitigation and roundwood production projects, 29th edn. Springer, The Netherlands

Rindfuss RR, Walsh SJ, Turner BL, Fox J, Mishra V (2004) Developing a science of land change: challenges and methodological issues. Proc Natl Acad Sci U S A 101:13976–13981. doi:10.1073/pnas.0401545101

Roder W, Phengchanh S, Keoboulapha B (1995) Relationships between soil, fallow period, weeds and rice yield in slash-and-burn systems of Laos. Plant Soil 176:27–36

Saatchi SS, Houghton R, Dos Santos Alvalá RC, Soares JV, Yu Y (2007) Distribution of aboveground live biomass in the Amazon basin. Glob Chang Biol 13:816–837. doi:10.1111/j.1365-2486.2007.01323.x

Sathaye J, Andrasko K (2007) Land use change and forestry climate project regional baselines: a review. Mitig Adapt Strateg Glob Chang 12:971–1000. doi:10.1007/s11027-006-9061-6

Schmook B (2010) Shifting maize cultivation and secondary vegetation in the Southern Yucatán: successional forest impacts of temporal intensification. Reg Environ Chang 10:233–246. doi:10.1007/s10113-010-0128-2

Serneels S, Said MY, Lambin EF (2001) Land cover changes around a major east African wildlife reserve: The Mara Ecosystem (Kenya). Int J Remote Sens 22:3397–3420. doi:10.1080/01431160152609236

Smith P, Davies C, Ogle S, Zanchi G, Bellarby J, Bird N, Boddey RM, McNamara NP, Powlson D, Cowie A, Noordwijk M, Davis SC, Richter DDB, Kryzanowski L, Wijk MT, Stuart J, Kirton A, Eggar D, Newton-Cross G, Adhya TK, Braimoh AK (2012) Towards an integrated global framework to assess the impacts of land use and management change on soil carbon: current capability and future vision. Glob Chang Biol 18:2089–2101. doi:10.1111/j.1365-2486.2012.02689.x

Smits PC, Dellepiane SG, Schowengerdt RA (1999) Quality assessment of image classification algorithms for land-cover mapping: a review and a proposal for a cost-based approach. Int J Remote Sens 20:1461–1486

Southworth J, Munroe D, Nagendra H (2004) Land cover change and landscape fragmentation—comparing the utility of continuous and discrete analyses for a western Honduras region. Agric Ecosyst Environ 101:185–205. doi:10.1016/j.agee.2003.09.011

Sydenstricker-Neto J, Parmenter AW, DeGloria S (2004) Participatory reference data collection methods for accuracy assessment of land-cover change maps. In: Lunetta RS, Lyon JG (eds) Remote sensing and GIS accuracy assessment. CRC Press, Boca Raton, pp 75–90. doi:10.1201/9780203497586.ch6

Tomich T, Van Noordwijk M, Budidarsono S, Gillison A, Kusumanto T, Murdiyarso D, Stolle F, Fagi AM (2001) Agricultural intensification, deforestation and the environment: assessing trade-offs in Sumatra, Indonesia. In: Lee DR, Barrett CB (eds) Trade-offs or synergies? Agricultural intensification, economic development, and the environment. CAB International, Wallingford, pp 221–244

UNFCCC/CCNUCC (2007) A/R methodological tool: combined tool to identify the baseline scenario and demonstrate additionality in A/R CDM project activities (Version 01) (No. EB 35 Report Annex 19) United Nations Framework Convention on Climate Change Clean Development Mechanism. https://cdm.unfccc.int/Reference/tools/index.html. Accessed 14 March 2015

UNFCCC/CCNUCC (2009) A/R methodological tool: calculation of the number of sample plots for measurements within A/R CDM project activities (Version 02) (No. EB 58 Report Annex 15) United Nations Framework Convention on Climate Change Clean Development Mechanism. https://cdm.unfccc.int/Reference/tools/index.html. Accessed 14 March 2015

Van Oort PJ (2007) Interpreting the change detection error matrix. Remote Sens Environ 108:1–8. doi:10.1016/j.rse.2006.10.012

VCS Association (2010) VM0006: methodology for carbon accounting in project activities that reduce emissions from mosaic deforestation and degradation (Version 1.0) Verified Carbon Standard, Washington, DC. http://www.v-c-s.org/methodologies/methodology-carbon-accounting-mosaic-and-landscape-scale-redd-projects-v21. Accessed 14 March 2015

VCS Association (2012) VT0001: tool for the demonstration and assessment of additionality in VCS Agriculture, Forestry and Other Land Use (AFOLU) project activities (Version 3.0) Verified Carbon Standard, Washington, DC. http://www.v-c-s.org/methodologies/tool-demonstration-and-assessment-additionality-vcs-agriculture-forestry-and-other. Accessed 14 March 2015

VCS Association (2014) Carbon accounting for grouped mosaic and landscape-scale REDD projects VM0006: methodology for carbon accounting in project activities that reduce emissions from mosaic deforestation and degradation (Version 2.1) Verified Carbon Standard, Washington, DC. http://www.v-c-s.org/methodologies/methodology-carbon-accounting-mosaic-and-landscape-scale-redd-projects-v21. Accessed 14 March 2015

Verburg PH, van de Steeg J, Veldkamp A, Willemen L (2009) From land cover change to land function dynamics: a major challenge to improve land characterization. J Environ Manage 90:1327–1335. doi:10.1016/j.jenvman.2008.08.005

Vinciková H, Hais M, Brom J, Procházka J, Pecharová E (2010) Use of remote sensing methods in studying agricultural landscapes—a review. J Landsc Stud 3:53–63

Chapter 4
Quantifying Greenhouse Gas Emissions from Managed and Natural Soils

Klaus Butterbach-Bahl, Björn Ole Sander, David Pelster, and Eugenio Díaz-Pinés

Abstract Standard methods for quantifying GHG emissions from soils tend to use either micrometeorological or chamber-based measurement approaches. The latter is the most widely used technique, since it can be applied at low costs and without power supply at remote sites to allow measurement of GHG exchanges between soils and the atmosphere for field trials. Instrumentation for micrometeorological measurements meanwhile is costly, requires power supply and a minimum of 1 ha homogeneous, flat terrain. In this chapter therefore we mainly discuss the closed chamber methodology for quantifying soil GHG fluxes. We provide detailed guidance on existing measurement protocols and make recommendations for selecting field sites, performing the measurements and strategies to overcome spatial variability of fluxes, and provide knowledge on potential sources of errors that should be avoided. As a specific example for chamber-based GHG measurements we discuss sampling and measurement strategies for GHG emissions from rice paddies.

K. Butterbach-Bahl (✉)
International Livestock Research Institute (ILRI),
Old Naivasha Rd., P.O. Box 30709, Nairobi, Kenya

Karlsruhe Institute of Technology, Institute of Meteorology and Climate Research,
Atmospheric Environmental Research (IMK-IFU),
Kreuzeckbahnstr. 19, Garmisch-Partenkirchen, Germany
e-mail: K.Butterbach-Bahl@cgiar.org

B.O. Sander
International Rice Research Institute (IRRI), Los Baños, Philippines

D. Pelster
International Livestock Research Institute (ILRI), Nairobi, Kenya

E. Díaz-Pinés
Karlsruhe Institute of Technology, Institute of Meteorology and Climate Research,
Atmospheric Environmental Research (IMK-IFU), Garmisch-Partenkirchen, Germany

T.S. Rosenstock et al. (eds.), *Methods for Measuring Greenhouse Gas Balances and Evaluating Mitigation Options in Smallholder Agriculture*,
DOI 10.1007/978-3-319-29794-1_4

4.1 Introduction

Microbial processes in soils, sediments, and organic wastes such as manure are a major source of atmospheric greenhouse gases (GHG). These processes create spatially as well as temporally heterogeneous sources or sinks. Consequently, a thorough understanding of the underlying processes and a quantification of spatiotemporal dynamics of sinks and sources are the bases for (a) developing GHG inventories at global, national, and regional scales, (b) identifying regional hotspots and (c) developing strategies for mitigating GHG emissions from terrestrial, specifically agricultural systems.

At the ecosystem scale, biosphere–atmosphere fluxes of CO_2, CH_4, and N_2O are bi-directional, i.e., what is observed is a net flux of production and consumption processes (e.g., CO_2: photosynthesis and autotrophic and heterotrophic respiration; CH_4: methanogenesis and methane oxidation; N_2O: nitrification and de-nitrification as source processes and de-nitrification as a sink process). The same is true for soil–atmosphere exchange processes, though, with regard to CO_2, often only respiratory fluxes are measured.

Approximately 2/3 of all N_2O emissions are linked to soil and manure management (Fowler et al. 2009; IPCC 2013). For CH_4 as well, soils and organic wastes strongly influence atmospheric CH_4 concentrations. It is estimated that wetland and paddy soils represent approximately 1/3 of all sources for atmospheric CH_4 (Fowler et al. 2009). On the other hand, well-aerated soils of natural and semi-natural ecosystems—and to a lesser extent soils of agroecosystems—are sinks for atmospheric CH_4, removing approximately 20–45 Tg yr^{-1} of CH_4 from the atmosphere (Dutaur and Verchot 2007), which corresponds to approximately 6–8 % of all sinks for atmospheric CH_4 (Fowler et al. 2009). For CO_2, soils are a major source due to autotrophic (plant root) and heterotrophic (microbial and soil fauna breakdown of organic matter) respiration. However, at the ecosystem scale, soils can act as net sinks as well as sources for CO_2, since at this scale plant primary production (CO_2 fixation from the atmosphere by photosynthesis), litter input to soils as well as respiratory fluxes are considered. It is well established that soils to a depth of 1 m globally store approximately three times the amount of carbon currently found in the atmosphere (Batjes 1996; IPCC 2013). Thus, land use and land management changes, as well as changes in climate affect plant primary production and fluxes of litter to the soil and soil organic matter mineralization dynamics. This can either result in a mobilization of soil C and N stocks, or, with adequate management, turn soils into C sinks. The latter is an essential process for removal of atmospheric CO_2 and climate protection and has been called the "recarbonization" of our terrestrial ecosystems (Lal 2009).

Due to the mostly microbiological origin of soil, sediment, and organic waste GHG emissions, changes in environmental conditions directly affect the exchange of GHG between terrestrial systems and the atmosphere (Butterbach-Bahl and Dannenmann 2011). Changes in temperature affect enzyme activities, while changes in redox conditions—as influenced by soil aeration fluctuations as a consequence of

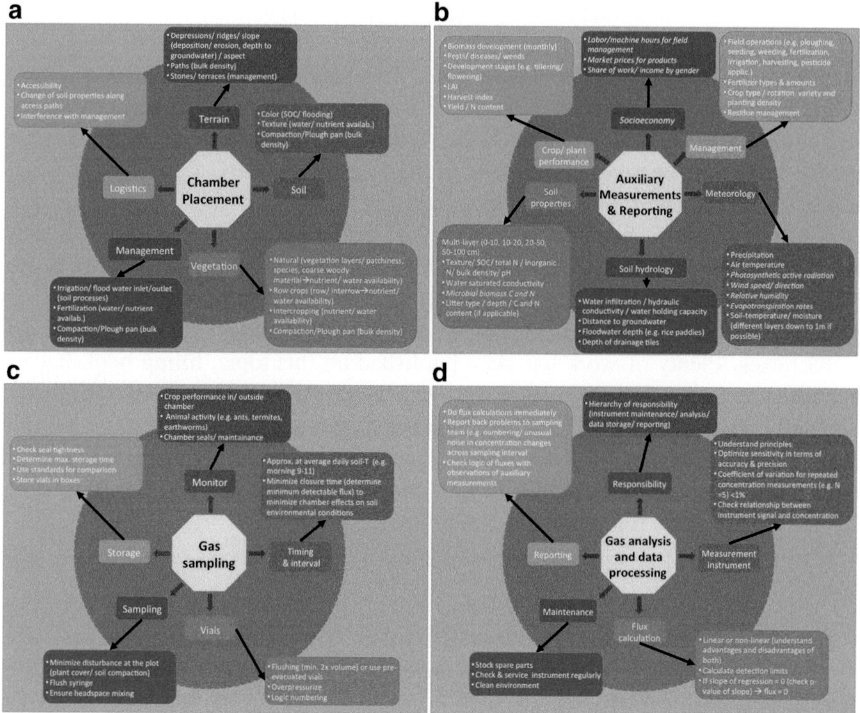

Fig. 4.1 General recommendations for chamber placement, gas sampling, gas concentration measurements, and measurement of auxiliary parameters for static chamber soil GHG flux measurements. (*Note*: text in *italic* are additional measurements/parameters which might be worthwhile to observe)

changes in soil moisture—can favor sequentially different microbial processes. For example, field irrigation and flooding as a standard management for rice paddies results in anaerobic soil conditions, thereby slowing down and stopping aerobic decomposition processes, while sequentially initializing a series of microbial processes that use elements and compounds other than oxygen as an electron acceptor: first NO_3^- (denitrification), followed by SO_4^- and Fe^{3+} and $Mn^{3+/4+}$ reduction, before finally CH_4 is produced as a product of organic matter degradation under strictly anaerobic conditions by methanogens (Conrad 1996).

Environmental conditions not only change naturally across days, seasons, and years as a consequence of diurnal and seasonal temperature and rainfall regimes, but also due to management of agricultural (forest with regard to plantations) land, as was explained above with the example of flooding of paddy fields. Changes in environmental conditions affect the activity of the microbial community as well as that of plants, and consequently, the associated GHG production and consumption processes. Thus, GHG emissions from soils show a rather pronounced temporal variability on short (diurnal) and longer (days to weeks and years) timescales (e.g., Luo et al. 2012). Moreover, environmental conditions also vary spatially because soil conditions, plant cover, land management and thus, nutrient availability, soil aeration and microbial community composition, also change across micro- (e.g., soil matrix) to landscape and continental scales. As a result, GHG fluxes also vary considerably across spatial scales, making it necessary to develop a solid sampling strategy to target measurement sites, i.e., determine which sites are representative for the landscape one would like to work in, to estimate GHG fluxes and develop strategies to mitigate those emissions. Targeting (Chap. 2 of these guidelines) is a cornerstone to allow meaningful upscaling to landscape and higher spatial scales. But targeting already starts at the measurement site, since decisions have to be made about where (and when) to place chambers for flux measurements (Fig. 4.1a).

This chapter does not aim to provide a cookbook of how to measure soil and GHG fluxes. Plenty of work has been published on this topic, filling bookshelves and libraries (see e.g., Table 4.1). Here, we provide guidance to the relevant literature and highlight potential problems that might come up when designing a GHG measurement program (Fig. 4.1) rather than explain the sampling procedures in detail. We also provide examples of how to overcome problems in the context of GHG measurements for smallholder systems.

4.2 What Technique Is Most Suitable for Measuring Biosphere–Atmosphere Exchange Processes of GHGs?

The two most commonly used techniques for measuring fluxes between terrestrial ecosystems and the atmosphere are: (a) enclosure-based (chamber) measurements (manual or automated) and (b) micrometeorological measurements (e.g., eddy covariance or gradient methods), or a combination of both (Denmead 2008). The choice of the measurement technique itself is largely driven by resource investment, demand, and by the research question.

4.2.1 Micrometeorological Measurements

Use of micrometeorological techniques requires homogenous fields with a significant fetch (>1 ha) that should not be influenced by buildings, trees, slopes, etc. Land use, land management, vegetation, and soil properties should be homogeneous for the direct fetch area, but also for the wider area. Typically these techniques are applied in flat terrain with large, homogeneous land use, such as pasture, grassland, maize, or wheat monocrops, forests, or tree plantations. Capital costs of micrometeorological measurements of GHG fluxes are high, since the required sensors (3D wind field, fast-response gas analyzers) plus auxiliary instruments (meteorological station, mast, etc.) for flux measurements at one site, cost around 60,000–80,000 USD for CO_2 and energy fluxes alone. Adding other components, such as CH_4 (open path sensors are available) and N_2O (requiring laser spectroscopy instruments),

Table 4.1 Literature overview on published protocols and recommendations for soil–atmosphere GHG measurements with emphasis on the static chamber methodology

Topic	Specialization	Methods	Highlight	References
General overviews on methodology				
Measuring biosphere-atmosphere exchange of CH_4 and N_2O	Overview on measuring techniques	Theoretical and practical information on measurements	Very good overview on measuring techniques	Denmead (2008)[a]
Rice paddies/wetlands/uplands	CH_4 flux measurement methods	Overview of techniques	Overview on CH_4 measuring techniques (micromet, chambers)	Schütz and Seiler (1992)
CH_4 and N_2O fluxes from livestock systems	Review	Description of approaches and underlying mechanisms	Review, incl. processes, methodology	Kebreab et al. (2006)
Overview on measuring techniques with focus on static chambers flux measurements	Overview on techniques	Provides practical guidance on measuring soil GHG fluxes	Overview on methodologies and shortcomings	Butterbach-Bahl et al. (2011)
Quality assurance for static chamber measurements	Quality assurance	Minimum set of criteria for static chamber design and deployment methodology	Confidence in the absolute flux values reported in about 60 % of the studies was estimated to be very low due to poor methodologies or incomplete reporting	Rochette and Eriksen-Hamel (2007)[a]
Micrometeorological measurements of N_2O, CO_2, CH_4	Micrometeorology	Description of procedures	Theory and application of micrometeorological measurements of GHG fluxes from agricultural fields	Pattey et al. (2006)

(continued)

Table 4.1 (continued)

Topic	Specialization	Methods	Highlight	References
Chamber measurement protocols				
Protocol for soil N_2O flux measurements	Detailed description of all steps for soil gas flux measurements	Static chamber, focus on N_2O	Detailed step by step description of procedures	De Klein and Harvey (2012)[a]
Protocol for measurements of N_2O and CH_4 fluxes from agricultural sources	Wide range of different techniques	Good overview about micrometeorological and chamber techniques, incl. techniques to measure CH_4 emissions from ruminants	Standard textbook on methods to measure agricultural GHG fluxes for reference	IAEA (1992)
Protocol for chamber measurements	Focus on chamber-based flux measurements of N_2O, CH_4, CO_2	Provides overview on calculations and practical recommendations for measurements	Standard protocol for the USDA-ARS GRACEnet project	Parkin and Venterea (2010)[a]
Protocol for chamber measurements in rice paddies	CH_4 fluxes from rice paddies	Sampling times and dates across the rice growing season	Simplified measuring protocol for CH_4 fluxes from rice paddies to minimize the number of measurements	Buendia et al. (1998)
Protocol for soil N_2O flux measurements	Description of protocols for N_2O measurements	Overview on static chamber methodology with focus on N_2O	Discusses potential errors when installing static chambers and provides minimum requirements for using these chambers	Rochette (2011)
Common practices for manual GHG sampling	Literature review on protocols as being practiced	Static closed chamber	Most widely used methodological features of manual GHG sampling identified	Sander et al. (2014b)
Protocol for gas pooling technique for static chamber measurements	Gas pooling technique	Overcoming spatial heterogeneity with static chambers	Pooling of gas samples across individual chambers is an acceptable approach to integrate spatial heterogeneity	Arias-Navarro et al. (2013)

Topic	Specialization	Methods	Highlight	References
Flux calculation for static chamber technique				
Flux calculation	Non-linear versus linear calculation methods for soil N_2O fluxes	Static chamber	Linear calculation schemes are likely more robust to relative differences in fluxes	Venterea et al. (2009)[a]
Flux calculation	Diffusion model	Static chamber	Common measurement practices and flux calculations underestimate emission rates by 15–25 % under most circumstances; error dependent on chamber height, soil–air porosity, and flux calculation method	Livingston et al. (2005)[a]
Flux calculation	Flux correction for static chamber measurements of N_2O and CO_2 fluxes	Static chambers	Correction scheme for estimating the magnitude of flux underestimation arising from chamber deployment	Venterea (2010)
Flux calculation	Flux correction	Static chambers	The systematic error due to linear regression is of the same order as the estimated uncertainty due to temporal variation	Kroon et al. (2008)
Flux calculation	Flux correction	Static chamber	Linear versus non-linear, provides link to free R software download for flux calculation	Pedersen et al. (2010)[a]
Flux calculation	Flux correction	Static chambers	Significant underestimation of soil CO_2 flux strength if linear regression is applied	Kutzbach et al. (2007)

(continued)

Table 4.1 (continued)

Topic	Specialization	Methods	Highlight	References
Flux calculation	Theoretical evaluation	Static chamber	Measurement and simulation of measuring errors	Hutchinson and Rochette (2003)
Static chamber N_2O fluxes	Headspace N_2O increase	Changes in soil gas concentrations upon chamber closure	Increased headspace concentration of N_2O reduced effective efflux of N_2O from the soil	Conen and Smith (2000)
Chamber design and comparison of methods				
Comparison of chamber designs and flux calculation	Linear versus non-linear flux calculation	Static chamber comparison	Increasing chamber height, area, and volume significantly reduces flux underestimation	Pihlatie et al. (2012)
Chamber measurements of N_2O fluxes from soils	Focus on soil N_2O fluxes	Closed and dynamic chambers	Comparison of different chamber types (sizes) with eddy covariance fluxes	Smith et al. (1996)
Static chamber design	Soil N_2O fluxes	Recommendations for chamber and vent design and flux calculation method	Vent dimension affects N_2O fluxes; one of the first papers on chamber design, flux calculations, and venting	Hutchinson and Mosier (1981)
Venting of static chambers				
Venting of closed chambers	Comparison of vented versus non-vented chambers	Closed chamber N_2O fluxes	Venting can create larger errors than the ones it is supposed to overcome	Conen and Smith 1998
Venting of closed chambers	Comparison of vented versus non-vented chambers	Closed chamber CO_2 fluxes for forest soils	Increases of CO_2 fluxes exceeding a factor of 2 in response to wind events for vented chambers	Bain et al. (2005)

Topic	Specialization	Methods	Highlight	References
Venting of closed chambers	Vent design	Closed chambers	Presenting a new vent design to avoid overestimation of CO_2 fluxes under windy conditions due to the Venturi effect	Xu et al. (2006)
Venting of closed chambers	Vent design and seals	Closed chambers	Discussion on the necessity of vents and of appropriate flux calculation	Hutchinson and Livingston (2001)
Chambers and small-scale variability of fluxes				
Chambers and small-scale heterogeneity of soil properties	Effect of soil physical characteristics on fluxes	Flux calculation methods in relation to soil properties	Reiterates effects of non-steady soil conditions on errors while measuring fluxes with chambers	Venterea and Baker (2008)
Static chamber measurements of soil CO_2 fluxes	Spatial heterogeneity, flux calculation	Frequency of sampling and the number of chambers for overcoming spatial heterogeneity	Means of eight randomly chosen flux measurements from a population of 36 measurements made with 300 cm^2 diameter chamber were within 25 % of full population mean 98 % of the time and were within 10 % of the full population mean 70 % of the time	Davidson et al. (2002)
Protocol for gas pooling technique for static chamber measurements	Gas pooling technique	Overcoming spatial heterogeneity with static chambers	Pooling of gas samples across individual chambers is an acceptable approach to integrate spatial heterogeneity	Arias-Navarro et al. (2013)

(continued)

Table 4.1 (continued)

Topic	Specialization	Methods	Highlight	References
Timing of measurements, sampling frequency, and cumulative fluxes				
Sampling frequency and N_2O flux estimates	Comparison of autochambers with replicated manual chambers	Evaluating the importance of sampling time	Autochambers are useful if significant diurnal fluctuations in temperature are expected and for better quantifying fertilization emission pulses	Smith and Dobbie (2001)
Sampling frequency and N_2O flux estimates	Automated measuring system	Effect of sampling frequency on estimates of cumulative fluxes	Sampling once every 21 days yielded estimates within −40 % to +60 % of the actual cumulative flux	Parkin (2008)
Sampling frequency and N_2O flux estimates	Automated measuring system	Evaluation of effects of sampling frequency on flux estimates	Low frequency measurements might lead to annual estimates which differ widely from continuous, automated flux measurements (e.g., 1 week = −5 to +20 %)	Liu et al. (2010)
Static chamber measurements	Comparison of flux estimates by automated and manual chambers	Chamber effects on soil environmental conditions	Seasonal cumulative N_2O and CH_4 fluxes as measured by manual chambers on daily basis were overestimated 18 % and 31 %, since diurnal variation in fluxes were not accounted for. On the other side, automated chambers reduced soil moisture. To avoid this, change of chamber positions is recommended	Yao et al. (2009)

Topic	Specialization	Methods	Highlight	References
CH_4 and N_2O flux measurements from manure slurry storage system	Comparison of continuous and non-continuous flux measurements	Recommendations of sampling intervals and timing of measurements	For CH_4, sampling between 1800 and 0800 h at intervals <7 days yielded ±10 % deviation for N_2O was 50 % when sampling at 2000 h	Wood et al. (2013)

[a]Recommended reading

requires a significant additional investment in instruments, starting from 30,000 to 40,000 USD per gas. Energy supply for the instruments (if not only focused on open path CO_2/H_2O/CH_4 technology) is another constraint that should be considered. The two most prominent global networks for multi-site and multi-species observations of biosphere–atmosphere-exchange of GHGs using micrometeorological methodologies are the National Ecological Observatory Network (NEON) in the USA (http://neoninc.org/) and the Integrated Carbon Observation Network (ICOS) in Europe (http://www.icos-infrastructure.eu/?q=node/17). Both networks offer information, processing tools for calculating fluxes and experts for providing support for designing, establishing, and running micrometeorological measurements.

Micrometeorological techniques for assessing GHG exchange are not recommended for smallholder systems due to the complexity of land uses and land management, small-scale gradients in soil fertility, and complex crop rotations with intercropping (Chikowo et al. 2014).

Some literature for a first reading on micrometeorological techniques is listed in Table 4.1.

4.2.2 *Chamber Measurements*

This technique allows measurements of GHG fluxes at fine scales, with chambers usually covering soil areas <1 m^2, and are thus much better suited for smallholder farming systems. They can be operated manually or automatically (Breuer et al. 2000). Chamber measurements are rather simple and therefore the most common approach for GHG measurements since they allow gas samples to be stored for future analysis and, with the exception of automated systems, they do not require power supply at the site. In contrast with micrometeorological approaches, chambers are suitable for exploring treatment effects (e.g., fertilizer and crop trials) or effects of land use, land cover, or topography on GHG exchange. However, care must be used in order to obtain accurate data, since installation of the chamber disturbs environmental conditions and measured fluxes might not necessarily reflect fluxes at adjacent sites if some precautions are not considered (see Sect. 5.2.1 below).

There are two types of chambers: dynamic and static chambers. For dynamic chambers the headspace air is exchanged at a high rate (>1–2 times the chamber's volume per minute) and fluxes are calculated from the difference in gas concentrations at the inlet and outlet of the chambers multiplied by the gas volume flux, thereby considering the area which is covered by the chamber (Butterbach-Bahl et al. 1997a, b). Static chambers are gas-tight, without forced exchange of the headspace gas volume, and are usually vented to allow pressure equalization between the chamber's headspace and the ambient air pressure (e.g., Xu et al. 2006). The volume of the "vent tube" should be greater than the gas volume taken at each sampling time.

Two situations call for using dynamic chambers: first, when measuring reactive gas fluxes such as soil NO emissions, and when there is a need to minimize the bias of changes in headspace air concentrations on the flux (Butterbach-Bahl et al. 1997a, b). The second point is important, as significant deviations of chamber headspace gas concentrations from ambient air concentrations affect the exchange process between soils and the atmosphere itself, since the flux at the soil–atmosphere interface is the result of simultaneous production and consumption processes. For example, if N_2O concentration in the chamber headspace is much higher than atmospheric concentrations, microbial consumption processes are stimulated. Moreover, since emissions are mainly driven by diffusion and gas concentration gradients, significant increases/decreases in headspace concentrations of the gas of interest will slow down/accelerate the diffusive flux. Both mechanisms finally result in a deviation of the flux magnitude from undisturbed conditions (Hutchinson and Mosier 1981). It is important to be aware of this, though for practical reasons it is partly unavoidable because the precision of the analytical instruments used for gas flux measurements, such as electron capture detectors (ECDs) and gas chromatography, is insufficient to allow for dynamic chamber measurements. However, there are methods to cope with this problem, such as using non-linear instead of linear models to calculate fluxes as measured with static chamber technique (e.g., Kroon et al. 2008; Table 4.1), using quantum cascade lasers (QCLs) in the field (fast box; Hensen et al. 2006) and in general by minimizing chamber closure time as much as possible. Chamber closure time is dictated not only by the magnitude of the gas flux but also by the chamber height. Therefore, in agricultural systems where plants need to be included for representative measurements, it is suggested to use chambers which can be extended by sections according to plant growth (Barton et al. 2008).

Static chambers are usually mounted on a frame which should be inserted (approximately 0.02–0.15 m) at least a week before first flux measurements to overcome initial disturbances of soil environmental conditions due to the insertion of the frame. Once the chamber is closed gas-tight on the frame, headspace concentrations start to change, either increasing if the soil is a net source (e.g., for CO_2—Fig. 4.2), or decreasing if the soil is functioning as a net sink (e.g., CH_4 uptake by upland soils). For accurate calculation of gas flux, a minimum of four gas samples from the chamber headspace across the sampling interval (e.g., 0, 10, 20, 30 min following closure) is recommended (Rochette 2011).

Gas flux measurements with static and dynamic chambers have been described extensively and Table 4.1 provides an overview of recommended literature, while Fig. 4.1 indicates important considerations when using chamber methodology. Static chambers can not only be used for measurement of soil N_2O and CH_4 and CO_2 respiratory fluxes, but also for measuring net ecosystem exchange of carbon dioxide. The latter requires the use of transparent chambers and consideration of corrections for photosynthetically active radiation and temperature inside and outside the chamber (Wang et al. 2013).

Chambers and Changes in Environmental Conditions

Closing a chamber gas-tight from the surrounding environment immediately affects a number of boundary conditions. The pressure inside the chamber might differ from outside, because when chambers are gas-tight and exposed to sunlight, the temperature of the headspace air increases so that air pressure inside in the chamber

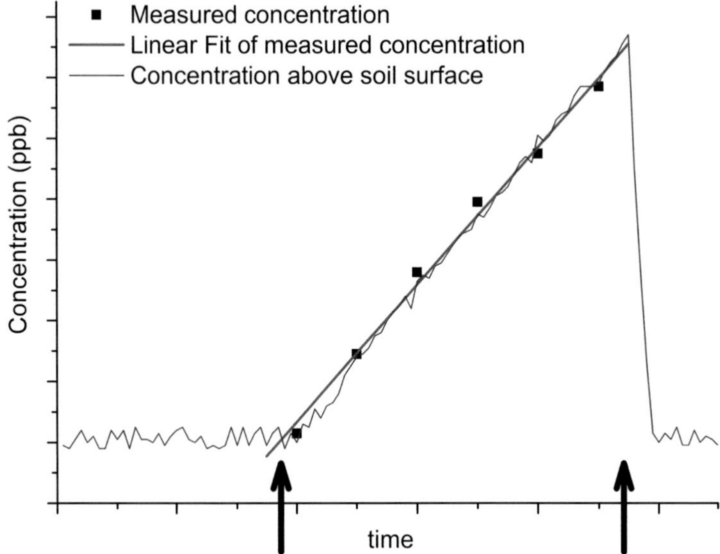

Fig. 4.2 Theoretical evolution of the concentration of a gas being emitted from the soil upon use of a static chamber. Concentration of the gas above the soil surface (*black line*) remains at a relatively constant level; at the moment when the chamber is closed (*left arrow*), the concentration in its headspace begins to rise. Along the closing period of the chamber, several gas samples are taken (*black squares*) and subsequently the concentration is determined, e.g., by use of gas chromatography. Right after opening the chamber (*right arrow*) concentration above soil surface returns to atmospheric background levels. Soil GHG emissions are most commonly calculated from the linear increase of the headspace gas concentration during the chamber closing period (*red line*), the volume of the chamber, the area of the soil covered by the chamber, as well as air temperature, air pressure, and molecular weight of the molecule under investigation (see e.g., Butterbach-Bahl et al. 2011). It should be noted that changes in gas concentration upon chamber closure can significantly deviate from linearity, showing, e.g., saturation effects. In all cases it should be tested if non-linear flux calculation methods do not fit the better observed changes in chamber headspace concentrations with time (see e.g., Pedersen et al. 2010)

increases too. Both factors affect the gas exchange between the soil and the air. Thus, chambers should be heat insulated and opaque (except for the determination of net ecosystem respiration; see Zheng et al. 2008a, b) and a vent should be used (see Hutchinson and Livingston 2001) to equilibrate pressure differences between ambient and headspace air. Upon chamber closure of transparent non-insulated chambers exposed to direct sunlight, headspace temperature might increase by 10–20 °C within 20 min. Insulated chambers will also show a slight increase in soil headspace temperature. This affects microbial as well as plant respiratory activity. Therefore, minimizing closure times is necessary not only to minimize the effects of changing headspace gas concentrations on diffusive fluxes as described above, but to minimize temperature changes as well as (Table 4.1). One should therefore calculate the minimum flux that can be detected with the analytical instrument to be used and adjust the closure time accordingly. If possible, limit closure time to a

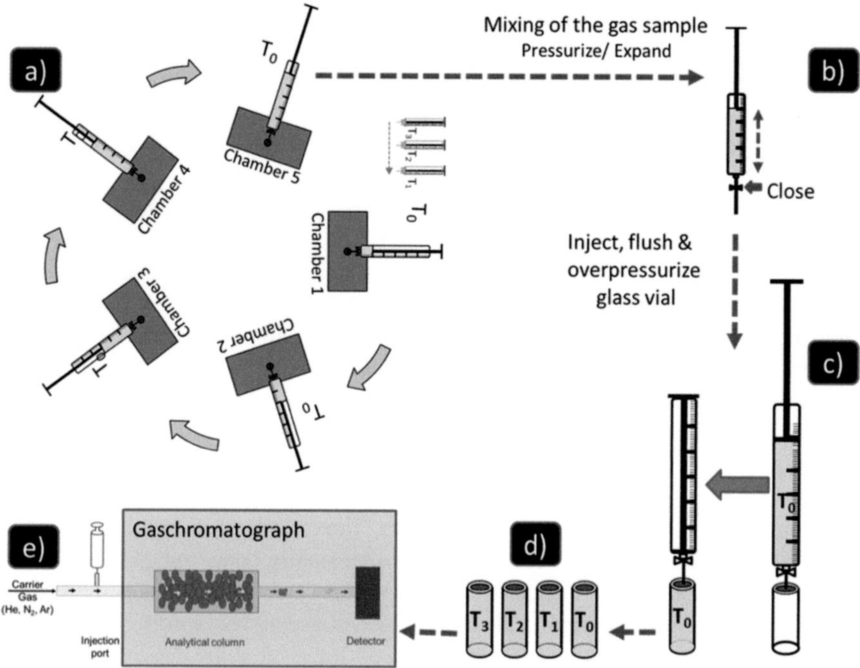

Fig. 4.3 The concept of gas pooling. (**a**) Gas pooling across chambers for a given sampling time, (**b**) gas sample mixing within the syringe, (**c**) transfer of the gas sample to a vial, (**d**) four vials for four sampling times and five chambers, (**e**) air sample analysis via gas chromatography (for further details see Arias-Navarro et al. 2013)

maximum of 30–45 min. If automated chamber systems are used, change positions weekly or at 2-week intervals to minimize effects on soil environmental conditions, in particular soil moisture. Chambers have been shown to reduce soil moisture even if they open automatically during rainfall (Yao et al. 2009).

Chambers and Spatial Variability of GHG Fluxes

Soil environmental conditions change on a small scale due to differences in (a) bulk density resulting from machine use or livestock grazing, (b) texture as a consequence of soil genesis, (c) management (rows, inter-rows, cropping), (d) temperature (plant shading), (e) soil moisture (e.g., groundwater distances or as an effect of texture differences), (f) soil organic carbon (heterogeneous distribution of harvest residues) or (g) rooting depth and distribution (with effects on soil microbial diversity, activity, and distribution) (see Fig. 4.1a). For example, urine or feces dropping by livestock on rangeland or manure application to cropland has been shown to increase spatial and temporal variability of fluxes, since at plot scale not every patch responds equally to increased availability of substrate for microbial N and C turnover processes due to small-scale differences in soil properties, soil environmental conditions, and microbial activity and diversity. Overcoming spatial variability effects on GHG fluxes is a major challenge, specifically for highly diverse smallholder systems. The problem can be addressed by proper sampling design (Fig. 4.1) (see e.g., Davidson et al. 2002) or by using the gas pooling technique (Arias-Navarro et al. 2013) (Fig. 4.3).

Proper sampling design in this context requires firstly that the landscape should be stratified into a number of separate categories. This stratification needs to include geophysical information as well as management activities. Also, in order to understand the drivers of the management decisions, it is critical to collect the political and socioeconomic climate of the various farms. The sampling approach can then concentrate measurement activities on emission hotspot and leverage points to capture heterogeneity and account for the diversity and complexity of farming activities (Rosenstock et al. 2013).

The gas pooling technique is similar to what is usually done for soil or water analyses. The principal idea of gas pooling is to generate a composite air sample out of the headspace of several chambers (Fig. 4.3). The chamber headspace is sampled at least four times across the closure period as is usually done, but gas samples at time 0, 10, 20, or 30 min are combined for several chambers of each individual sampling time (Arias-Navarro et al. 2013). As a consequence, information on the spatial variability is lost, but can be regained if on some sampling days, fluxes of the chambers are measured individually. This technique allows installation of a significantly higher number of chambers without increasing the amount of gas samples to be analyzed.

4.3 Measurement of GHG Fluxes in Rice Paddies

Due to its importance as a source for atmospheric CH_4 we specifically discuss measurement of GHG fluxes in rice paddies in more detail. Unlike other field crops, rice is usually grown in flooded fields. The standing water creates anaerobic conditions in the soil that allows growth of a certain class of microorganisms (*methanogenic archaea*) that use simple carbon compounds (e.g., CO_2 or acetate) as electron donors and produce methane in anaerobic respiration. Methane oxidation, on the other hand, does occur but only in the uppermost mm of flooded paddy soil or in the rhizosphere—due to radial O_2 losses of rice roots (Butterbach-Bahl et al. 1997a, b)—and during unflooded periods. Since methanogenic archaea are extremely sensitive to oxygen and immediately stop CH_4 production while stimulating CH_4 oxidation, drainage of rice fields is an attractive mitigation option.

Methane is the most important GHG in rice production systems and has some implications on the chamber design and sampling time. Nitrous oxide emissions are generally low in flooded fields but increase with drainage. However, this increase in N_2O emissions does not offset the mitigation effect that dry field conditions have on CH_4 emissions (Sander et al. 2014).

Table 4.2 Overview of recommended minimum requirements for closed chamber sampling in rice paddy and for measurements of field GHG fluxes from upland arable fields

	Minimum requirement/recommendation	
Feature	Rice paddy	Arable field
Chamber dimension	4 rice hills included, ≥0.16 m^2, >1 m height *or* extendable, chamber base ~20 cm high	Height 10–40 cm (flexible height if possible), insertion depth 5–20 cm, minimum area 0.04 m^2. Include plants as long as possible, consider row/inter-row effects
Chamber material	Reflective *or* white *and/or* insulated	Opaque, insulated (use transparent material only if NEE should be measured)
Chamber equipment	Thermometer, fan, sampling port, hole for irrigation water, vent	Thermometer, fan, vent
Frequency	Once per week *or* elaborated flexible schedule	Once per week, following the first 10 days after fertilization or re-wetting of dried soils if possible daily measurements
Length of measuring period	1 year	1 year
Spatial replicates	At least 3, possibly use gas pooling technique	At least 3, possibly use gas pooling technique
Time of day	At the time of approx. average daily soil temperature (often mid-morning). Record diurnal flux variation from time to time	Record diurnal flux variation
Closure time	As short as possible, as long as necessary, In hot environments 20–30 min, not more than 45 min	As short as possible, as long as necessary, In hot environments 20–30 min, not more than 45 min
Number of gas samples for flux calculation	≥4 per deployment	≥4 per deployment

These recommendations have been synthesized from prior chamber measurement protocols (see Table 4.1) and amended or modified on basis of expert judgments. For further details see also Fig. 4.1

Overall, requirements for GHG measurements in flooded rice production systems (dominated by CH_4 emissions) are partly different from measurement in upland systems, which has some important implications on the chamber design and general sampling procedure (Table 4.2).

4.3.1 Rice Chamber Design and General Procedure (See Also Table 4.2)

Methane that is produced in the soil has three different emission pathways to the atmosphere: (1) diffusion through the water layer, (2) ebullition (bubbling), and (3) transport through the aerenchyma of the rice plants. The largest share of emitted methane (up to 90 %) is in fact transported through the rice plant itself (Wassmann et al. 1996; Butterbach-Bahl et al. 1997a, b), which makes it indispensable to include rice plants into the closed chamber (→ chamber height >1 m). This also applies to any measurements of wetland GHG fluxes, since plant-mediated transport is of critical importance here as well. The chamber base (the part of the chamber that remains in the soil during the whole growing season) should be installed at least 1 day (better a week or more) before the start of the sampling campaign and should not be higher than ~20 cm (with 10 cm below and 10 cm above soil surface) in order to minimize an effect on plant growth. To account for variability within the field, each chamber should include at least 4 rice plants or 4 "hills" in a transplanted system and an area of average plant density in a seeded system, resulting in a chamber area of ≥ 0.16 m^2. Note that due to the flooded field conditions, the chamber base in rice systems should have holes (~2 cm above soil surface) to allow water exchange between the chamber inside and the field. This hole or holes must be closed before sampling in case irrigation water level falls and the hole(s) is above the water layer.

Movement in the wet paddy soil can potentially cause gas bubbles to evolve and impede undisturbed gas sampling. Therefore, installation of boardwalks in the field is highly recommended. Exposure to high air temperatures and high solar radiation often characterize rice paddies and so it is in especially crucial to ensure that the plants inside the chambers are not damaged by heat stress during sampling. Therefore, the chamber material should be reflective or white or the chamber should be equipped with proper insulation. Since the gas volume in the closed chamber changes due to temperature increase and samples being taken, chambers should have a vent to allow equilibration with outside air pressure.

4.3.2 Time of Day of Sampling

Methane emissions typically follow a distinct diurnal variation following changes in soil temperature (Neue et al. 1997), i.e., low emissions during night time that increase after sunrise, peak around noon to early afternoon and decrease again thereafter. Therefore the timing of gas sampling is of great importance in order to measure as close as possible to a time representing a daily average flux rather than at times leading to over or underestimation of fluxes. Minamikawa et al. (2012) found that methane fluxes around 10 a.m. were closest to the daily mean CH_4 flux in temperate regions. Similar assumptions are likely valid for tropical and subtropical regions. However, we recommend measuring region-specific diurnal emission patterns at least three times during the growing season of rice and based on the observed diurnal pattern to decide on the best sampling time. Alternatively, measuring diurnal soil temperature profiles at 5-cm depth can provide reasonable estimations of the time of day with mean methane emission because soil temperature and CH_4 flux are closely related.

4.3.3 Sampling Frequency

The precision of cumulative seasonal GHG emissions largely depends on the sampling frequency. Minamikawa et al. (2012) found that sampling once a week for flooded rice in temperate regions resulted in an accurate estimation of total emissions. Buendia et al. (1998) proposed a more flexible sampling schedule of 10-day intervals in the beginning of the growing season, 20-day intervals in the middle and 7-day intervals at the end of the season in tropical environments and came up with similarly accurate seasonal emission estimates.

It is important to note that more frequent sampling is necessary during dry periods of rice cultivation as methane emissions from paddy soils with a high clay content show a sharp peak when drainage is applied (Lu et al. 2000) and nitrous oxide emissions increase during dry periods (Jiao et al. 2006). In order to have complete flux information of an area, some gas samples should also be taken between two cropping seasons.

4.4 Analytical Instruments Used for Chamber Measurements

When using the static chamber approach, several analytical instruments can be used for determining GHG concentrations in the sample air, either directly in the field or, following storage of headspace gas samples in vials or gas-tight syringes, at a later time in the laboratory. The latter always requires that the gas-tightness of the vials/syringes is tested regularly.

4.4.1 Gas Chromatography

Instruments used for gas sample analysis rely on different operational principles. Gas chromatography (GC) is the most commonly used analytical technique when determining GHG concentrations in gas samples from chambers (e.g., Keller et al. 1986; Kiese and Butterbach-Bahl 2002; Kelliher et al. 2012). Usually, 1–3 mL of air sample is injected into the gas chromatograph and the different compounds are separated on an analytical column (e.g., Hayesep N for N_2O, 3 m, 1/8″) for detection with various detectors. For N_2O a ^{63}Ni Electron Capture Detector (ECD) is commonly used. The ECD should be operated at between 330 and 350 °C, since the N_2O sensitivity is highest and the cross-sensitivity to CO_2 is lowest in this range. However, there is still a cross-sensitivity to CO_2 if N_2 is used as sole carrier and purge gas (Zheng et al. 2008a, b; Wang et al. 2010). No cross-sensitivity exits if Argon/CH_4 is used as carrier gas or if the ECD cell is purged with a gas mixture of 5 % CO_2 in N_2 (Wang et al. 2010). Another possibility to eliminate the cross-sensitivity of N_2O and CO_2 is to use a pre-column filled with Ascarite (coated NaOH), which scrapes the CO_2 from the gas-stream. However, pre-columns need to be changed frequently (approximately 2-week intervals) due to saturation and capturing of air sample moisture.

Another critical point is that if gas chromatographs with ECD are used for concentration measurements, the signal to concentration ratio might deviate from a linear response if—in the case of N_2O—sample air concentrations are significantly >700 ppbv. Therefore, a check of the linearity of the signal to concentration ratio should be done for each instrument and gas under consideration.

For CH_4 a flame ionization detector (FID) is normally used and, if a methanizer is introduced before the detector, CO_2 can also be measured with a FID (or more standard: use of a thermal conductivity detector for CO_2).

4.4.2 Spectroscopic Methods

Spectroscopic methods are becoming more and more prominent for measuring GHG fluxes between soils and the atmosphere by static chamber technique. A specific example is photoacoustic spectroscopy (PAS), with instruments being miniaturized to make them suitable for direct field use, e.g., allowing direct measurements of changes in chamber headspace N_2O, CH_4, or CO_2 concentration with time following chamber closure (e.g., Leytem et al. 2011). PAS technique, as every spectroscopic method, is based on the principle that GHGs absorb light at a specific wavelength, here in the infrared spectra. The absorption is thereby directly linked to the concentration (Beer-Lambert law) and in the case of PAS, the absorption of the light or energy is converted into an acoustic signal, which is measured by a microphone. For chamber measurements in the field, the PAS instrument is usually connected to the chamber in a closed loop so that the air from the apparatus exhaust is returned to the chamber avoiding underpressure or dilution.

PAS instruments are becoming popular as an alternative to GC-technique due to portability, low maintenance, and ease-of-operation (Iqbal et al. 2012). In principle, commercially available PAS instruments, such as INNOVA (Lumasense Technologies) require a yearly calibration only and are "plug-and-play" instruments ready to be used in the field. However, because GHGs and water vapor have multiple absorption bands across the measuring spectra, such instruments are prone to interferences. Recently, Rosenstock et al. (2013) showed that for INNOVA instruments N_2O concentration measurements were non-linearly affected by water content and CO_2. Comparable results were already reported by Flechard et al. (2005), though only a few researchers have noted the problems that might be associated with the use of PAS. The manufacturers claim that the INNOVA software accounts for cross interferences, but corrections do not seem to work sufficiently while testing several instruments (Rosenstock et al. 2013). Furthermore, there is also evidence that ambient air temperature affects the electronics and thus, the reliability of measured GHG concentrations (Rosenstock et al. 2013), when using PAS under field conditions. Specifically for N_2O, measured concentrations varied up to 100 % depending on environmental conditions (Rosenstock et al. 2013). Also the precision and accuracy of CH_4 measurements seems to be rather low, with deviations in concentration of nearly 400 % for calibration gases (Rosenstock et al. 2013). As it stands now, it is advisable to question the use of INNOVA instruments for CH_4 as well as for N_2O measurements in particular by using the instrument for simultaneous measurements of multiple gas species.

Other techniques may include tunable diode lasers (TDL), quantum cascade lasers (QCL), Fourier transform infrared spectroscopy (FTIR) or cavity ring-down spectroscopy (CRDS). Instruments using these spectroscopic techniques usually operate under high vacuum and, thus, a continuous air flow through the instrument is required. Therefore, instruments need to be at the study site and physically connected to chambers. Though these instruments are still quite expensive (e.g., compared to GC) they are becoming more and more robust and suitable for field applications. However, a constant (use of UPS is suggested) main power supply is still needed and checks for cross-sensitivity should be a standard procedure in the laboratory.

4.4.3 Auxiliary Measurements

As described earlier in this chapter, spatiotemporal patterns of GHG fluxes are closely linked to changes in environmental conditions (see also Fig. 4.1). Therefore, GHG flux measurements are rather useless if environmental parameters such as soil and vegetation properties and management are not monitored at the same time, since these factors significantly affect fluxes. This necessarily also includes the quantification of soil C and N stocks, as for example application of animal manure to arable fields and rangeland has been shown to significantly increase soil carbon stocks (Maillard and Angers 2014), which need to be considered when calculating the GHG balance of a given system. Moreover, since GHG flux measurements are expensive and can't be repeated everywhere, models need to be developed, tested, and finally used for estimating fluxes at landscape, regional, and global scale as well as for exploring mitigation options at multi-year scales or for predicting climate change feedbacks on biosphere–atmosphere exchange processes. Comprehensive datasets, including both flux measurements and detailed information on soil and vegetation properties and management are prerequisites for model development and testing. Surprisingly such datasets are still scarce, because either flux measurements do not meet the required measuring standards or the needed auxiliary measurements and site information are not monitored or reported.

Since responsibilities for GHG flux and auxiliary measurements are often split between collaborators, there is a need to clarify personal responsibility of data provision prior to the start of measurements. Rochette and Eriksen-Hamel (2007) reviewed published N_2O flux data and developed a minimum set of criteria for chamber design and methodology. According to their evaluation of 365 studies, there was low to very low confidence in reported flux values in about 60 % of the studies due to poor methodologies or incomplete reporting. Thus, it is necessary to improve not only the quality of flux measurements, but also the reporting of soil and vegetation properties and management. See Fig. 4.1 for suggested variables for measurement.

4.5 Conclusions

Micrometeorological or chamber-based techniques can be used for the quantification of biosphere–atmosphere exchange processes of GHGs. In view of the diversity and patchiness of land uses and land management associated with smallholder agriculture, chamber-based methods, specifically the closed (static) chamber approach, is recommended. Overcoming spatial and temporal variability of fluxes remain an issue, and should be addressed by a well-designed sampling scheme including landscape targeting of measuring sites (see Rufino et al. this book), targeting of chamber placement at field and plot scale (Fig. 4.1), running of at least 3–5 replicates per plot to address small-scale variability (and possibly use of the gas pooling technique, Fig. 4.3), flux measurements in weekly intervals over a period of at least 1 year and detailed documentation of environmental conditions and field activities (Fig. 4.1). This will ensure that all data can finally be used for modeling and upscaling. Quality control and quality assurance remains an issue at all steps, also with regard to gas analytics. Probably the most efficient way for a researcher to familiarize him- or herself with gas flux measurement techniques is a longer stay with a recognized research group.

Acknowledgments This work was financed by the SAMPLES project of the Climate Change, Agriculture and Food Security (CCAFS) research program of CGIAR Centre and Research Programs. Eugenio Diaz-Pínés received additional funding from the EU-InGOS project.

References

Arias-Navarro C, Díaz-Pinés E, Kiese R, Rosenstock RS, Rufino MC, Stern D, Neufeldt H, Verchot LS, Butterbach-Bahl K (2013) Gas pooling: a sampling technique to overcome spatial heterogeneity of soil carbon dioxide and nitrous oxide fluxes. Soil Biol Biochem 67:20–23

Bain WG, Hutyra L, Patterson DC, Bright AV, Daube BC, Munger JW, Wofsy SC (2005) Wind-induced error in the measurement of soil respiration using closed dynamic chambers. Agric For Meteorol 131:225–232

Barton L, Kiese R, Gatter D, Butterbach-Bahl K, Buck R, Hinz C, Murphy DV (2008) Nitrous oxide emissions from a cropped soil in a semi-arid climate. Glob Chang Biol 14:177–192

Batjes NH (1996) Total carbon and nitrogen in the soils of the world. Eur J Soil Sci 47:151–163

Breuer L, Papen H, Butterbach-Bahl K (2000) N_2O-emission from tropical forest soils of Australia. J Geophys Res 105:26353–26378

Buendia LV, Neue HU, Wassmann R, Lantin RS, Javellana AM, Arah J, Wang Z, Wanfang L, Makarim AK, Corton TM, Charoensilp N (1998) An efficient sampling strategy for estimating methane emission from rice field. Chemosphere 36:395–407

Butterbach-Bahl K, Dannenmann M (2011) Denitrification and associated soil N_2O emissions due to agricultural activities in a changing climate. Curr Opin Environ Sustain 3:389–395

Butterbach-Bahl K, Papen H, Rennenberg H (1997a) Impact of gas transport through rice cultivars on methane emission from rice paddy fields. Plant Cell Environ 20(9):1175–1183

Butterbach-Bahl K, Gasche R, Breuer L, Papen H (1997b) Fluxes of NO and N_2O from temperate forest soils: impact of forest type, N deposition and of liming on the NO and N_2O emissions. Nutr Cycl Agroecosyst 48:79–90

Butterbach-Bahl K, Kiese R, Liu C (2011) Biosphere-atmosphere exchange of CH4 in terrestrial systems. In: Rosenzweig AC, Ragsdale SW (eds) Methods in enzymology, vol 49. Academic, New York, pp 271–287

Chikowo R, Zingore S, Snapp S, Johnston A (2014) Farm typologies, soil fertility variability and nutrient management in smallholder farming in Sub-Saharan Africa. Nutr Cycl Agroecosyst 100:1–18

Conen F, Smith KA (1998) A re-examination of closed chamber methods for the measurement of trace gas emissions from soils to the atmosphere. Eur J Soil Sci 49:701–707

Conen F, Smith KA (2000) An explanation of linear increases in gas concentration under closed chambers used to measure gas exchange between soil and the atmosphere. Eur J Soil Sci 51:111–117

Conrad R (1996) Soil microorganisms as controllers of atmospheric trace gases (H_2, CO, CH_4, OCS, N_2O, and NO). Microbiol Rev 60:609–640

Davidson EA, Savage K, Verchot LV, Navarro R (2002) Minimizing artifacts and biases in chamber-based measurements of soil respiration. Agric For Meteorol 113:21–37

De Klein CAM, Harvey MJ (2012) Nitrous oxide chamber methodology guidelines. Global Research Alliance, Ministry of Primary Industries, New Zealand, 146p. Available via DIALOG: www.globalresearchalliance.org/research/livestock/activities/nitrous-oxide-chamber-methodology-guidelines/

Denmead OT (2008) Approaches to measuring fluxes of methane and nitrous oxide between landscapes and the atmosphere. Plant Soil 309:5–24

Dutaur L, Verchot LV (2007) A global inventory of the soil CH_4 sink. Glob Biogeochem Cycles 21(4), GB4013

Flechard C, Neftel A, Jocher M, Ammann C, Fuhrer J (2005) Bi-directional soil/atmosphere N_2O exchange over two mown grassland systems with contrasting management practices. Glob Chang Biol 11:2114–2127

Fowler D, Pilegaard K, Sutton MA, Ambus P, Raivonen M, Duyzer J, Simpson D, Fagerli H, Fuzzi S, Schjoerring JK, Granier C, Neftel A, Isaksen ISA, Laj P, Maione M, Monks PS, Burkhardt J, Daemmgen U, Neirynck J, Personne E, Wichink-Kruit R, Butterbach-Bahl K, Flechard C, Tuovinen JP, Coyle M, Gerosa G, Loubet B, Altimir N, Gruenhage L, Ammann C, Cieslik S, Paoletti E, Mikkelsen TN, Ro-Poulsen H, Cellier P, Cape JN, Horvath L, Loreto F, Niinemets U, Palmer PI, Rinne J, Misztal P, Nemitz E, Nilsson D, Pryor S, Gallagher MW, Vesala T, Skiba U, Brüggemann N, Zechmeister-Boltenstern S, Williams J, O'Dowd C, Facchini MC, de Leeuw G, Flossman A, Chaumerliac N, Erisman JW (2009) Atmospheric composition change: ecosystems—atmosphere interactions. Atmos Environ 43:5193–5267

Giltrap DL, Berben P, Palmada T, Saggar S (2014) Understanding and analyzing spatial variability of nitrous oxide emissions from a grazed pasture. Agric Ecosyst Environ 186:1–10

Groffman PM, Butterbach-Bahl K, Fulweiler RW, Gold AJ, Morse JL, Stander EK, Tague C, Tonitto C, Vidon P (2009) Challenges to incorporating spatially and temporally explicit phenomena (hotspots and hot moments) in denitrification models. Biogeochemistry 93:49–77

Hensen A, Groot TT, Van den Bulk WCM, Vermeulen AT, Olesen JE, Schelde K (2006) Dairy farm CH_4 and N_2O emissions, from one square meter to the full farm. Agric Ecosyst Environ 112:146–152

Holst J, Liu C, Yao Z, Brüggemann N, Zheng X, Han X, Butterbach-Bahl K (2007) Importance of point sources on regional nitrous oxide fluxes in semi-arid steppe of Inner Mongolia, China. Plant and Soil 296:209–226

Hutchinson GL, Livingston GP (2001) Vents and seals in non-steady state chambers used for measuring gas exchange between soil and the atmosphere. Eur J Soil Sci 52:675–682

Hutchinson GL, Mosier AR (1981) Improved soil cover method for field measurement of nitrous oxide fluxes. Soil Sci Soc Am J 45:311–316

Hutchinson GL, Rochette P (2003) Non-flow through steady state chambers for measuring soil respiration: numerical evaluation of their performance. Soil Sci Soc Am J 67:166–180

International Atomic Energy Agency (IAEA) (1992) Manual on measurement of methane and nitrous oxide emissions from agriculture. In: Training Course Series No. 29, Vienna, pp 1–237

Iqbal J, Castellano MJ, Parkin TB (2012) Evaluation of photoacoustic infrared spectroscopy for simultaneous measurement of N_2O and CO_2 gas concentrations and fluxes at the soil surface. Glob Change Biol 19:327–336

Jiao Z, Hou A, Shi Y, Huang G, Wang Y, Chen X (2006) Water management influencing methane and nitrous oxide emissions from rice field in relation to soil redox and microbial community. Commun Soil Sci Plant Anal 37:1889–1903

Kebreab E, Clark K, Wagner-Riddle C, France J (2006) Methane and nitrous oxide emissions from Canadian animal agriculture: a review. Can J Anim Sci 86:135–158

Kroon PS, Hensen A, Van den Bulk WCM, Jongejan PAC, Vermeulen AT (2008) The importance of reducing the systematic error due to non-linearity in N_2O flux measurements by static chambers. Nutr Cycl Agroecosyst 82:175–186

Kutzbach L, Schneider J, Sachs T, Giebels M, Nykänen H, Shurpali NJ, Martikainen PJ, Alm J, Wilmking M (2007) CO2 flux determination by closed chamber methods can be seriously biased by inappropriate application of linear regression. Biogeosciences 4:1005–1025

Lal R (2009) Sequestering atmospheric carbon dioxide. Crit Rev Plant Sci 28:90–96

Leytem AB, Dungan RS, Bjorneberg DL, Koehn AC (2011) Emissions of ammonia, methane, carbon dioxide, and nitrous oxide from dairy cattle housing and manure management systems. J Environ Qual 40:1383–1394

Liu C, Zheng X, Zhou Z, Han S, Wang Y, Wang K, Liang W, Li M, Chen D, Yang Z (2010) Nitrous oxide and nitric oxide emissions from irrigated cotton field in Northern China. Plant Soil 332:123–134

Livingston GP, Hutchinson GL, Spartalian K (2005) Diffusion theory improves chamber-based measurements of trace gas emissions. Geophys Res Lett 32, L24817

Lu WF, Chen W, Duan BW, Guo WM, Lu Y, Lantin RS, Wassmann R, Neue HU (2000) Methane emissions and mitigation options in irrigated rice fields in southeast China. Nutr Cycl Agroecosyst 58:65–73

Luo GJ, Brüggemann N, Wolf B, Gasche R, Grote R, Butterbach-Bahl K (2012) Decadal variability of soil CO_2, NO, N_2O, and CH_4 fluxes at the Höglwald Forest, Germany. Biogeosciences 9:1741–1763

Luo J, Hoogendorn C, Van der Weerden T, Saggar S, De Klein C, Giltrap D, Rollo M, Rys G (2013) Nitrous oxide emissions from grazed hill in New Zealand. Agric Ecosyst Environ 181:58–68

Maillard É, Angers DA (2014) Animal manure application and soil organic carbon stocks: a meta-analysis. Glob Chang Biol 20:666–679

Minamikawa K, Yagi K, Tokida T, Sander BO, Wassmann R (2012) Appropriate frequency and time of day to measure methane emissions from an irrigated rice paddy in Japan using the manual closed chamber method. Greenhouse Gas Meas Manag 2(2–3):118–128

Neue HU, Wassmann R, Kludze HK, Wang B, Lantin RS (1997) Factors and processes controlling methane emissions from rice fields. Nutr Cycl Agroecosyst 49:111–117

Parkin TB (2008) Effect of sampling frequency on estimates of cumulative nitrous oxide emissions. J Environ Qual 37:1390–1395

Parkin TB, Venterea RT (2010) Sampling protocols. Chap 3 Chamber based trace gas flux measurements. In: Folett RF (ed) GRACEnet Sampling protocols, pp 3-1–3-39. www.ars.usda.gov/research/GRACEnet

Pattey E, Edwards G, Strachan IB, Desjardins RL, Kaharabata S, Wagner-Riddle C (2006) Towards standards for measuring greenhouse gas fluxes from agricultural fields using instrumented towers. Can J Soil Sci 86:373–400

Pedersen AR, Petersen SO, Schelde K (2010) A comprehensive approach to soil-atmosphere trace-gas flux estimation with static chambers. Eur J Soil Sci 61:889–902

Pihlatie M et al (2012) Comparison of static chambers to measure CH_4 emissions from soils. Agric For Meteorol 171–172:124–136

Rochette P (2011) Towards a standard non-steady-state chamber methodology for measuring soil N2O emissions. Anim Feed Sci Technol 166(167):141–146

Rochette P, Eriksen-Hamel NS (2007) Chamber measurements of soil nitrous oxide flux: are absolute values reliable? Soil Sci Soc Am J 72:331–342

Rosenstock TS, Diaz-Pines E, Zuazo P, Jordan G, Predotova M, Mutuo P, Abwanda S, Thiong'o M, Buerkert A, Rufino MC, Kiese R, Neufeldt H, Butterbach-Bahl K (2013) Accuracy and precision of photoacoustic spectroscopy not guaranteed. Glob Chang Biol 19:3565–3567

Sander BO, Wassmann R (2014) Common practices for manual greenhouse gas sampling in rice production: a literature study on sampling modalities of the closed chamber method. Greenhouse Gas Meas Manag 4(1):1–13

Sander B, Wassmann R, Siopongco JDLC (2014) Water-saving techniques: potential, adoption and empirical evidence for mitigating greenhouse gas emissions from rice production. In: Hoanh CT, Smakhtin V, Johnston T (eds) Climate change and agricultural water management in developing countries. CABI Climate Change Series. CABI, Wallingford

Schütz H, Seiler W (1992) Methane flux measurements: methods and results. In: Andreae MO, Schimel DS (eds) Exchange of trace gases between terrestrial ecosystems and the atmosphere. Wiley, Chichester, pp 209–228

Smith KA, Dobbie KE (2001) The impact of sampling frequency and sampling times on chamber based measurements of N_2O emissions from fertilized soils. Glob Chang Biol 7:933–945

Smith KA, Clayton H, McTaggart IP, Thomson PE, Arah JRM, Scott A, Goulding KWT, Monteith JL, Phillips VR (1996) The measurement of nitrous oxide emissions from soil by using chambers (and discussion). Philos Trans Phys Sci Eng 351:327–338

Venterea RT (2010) Simplified method for quantifying theoretical underestimation of chamber-based trace gas fluxes. J Environ Qual 39:126–135

Venterea RT, Baker JM (2008) Effects of soil physical non-uniformity on chamber-based gas flux estimates. Soil Sci Soc Am J 72:1410–1417

Venterea RT, Spokas KA, Baker JM (2009) Accuracy and precision analysis of chamber-based nitrous oxide gas flux estimates. Soil Sci Soc Am J 73:1087–1093

Wang Y, Wang Y, Ling H (2010) A new carrier gas type for accurate measurement of N_2O by GC-ECD. Adv Atmos Sci 27:1322–1330

Wang K, Liu C, Zheng X, Pihlatie M, Li B, Haapanala S, Vesala T, Liu H, Wang Y, Liu G, Hu F (2013) Comparison between eddy covariance and automatic chamber techniques for measuring net ecosystem exchange of carbon dioxide in cotton and wheat fields. Biogeosciences 10:6865–6877

Wassmann R, Neue HU, Alberto MCR, Lantin RS, Bueno C, Llenaresas D, Arah JRM, Papen H, Seiler W, Rennenberg H (1996) Fluxes and pools of methane in wetland rice soils with varying organic inputs. Environ Monit Assess 42:163–173

Wood JD, Gordon RJ, Wagner-Riddle C (2013) Biases in discrete CH_4 and N_2O sampling protocols associated with temporal variation of gas fluxes from manure storage systems. Agric For Meteorol 171–172:295–305

Xu L, Furtaw MD, Madsen RA, Garcia RL, Anderson DJ, McDermitt DK (2006) On maintaining pressure equilibrium between a soil CO_2 flux chamber and the ambient air. J Geophys Res 111:D08S10

Yao Z, Zheng X, Xie B, Liu C, Mei B, Dong H, Butterbach-Bahl K, Zhu J (2009) Comparison of manual and automated chambers for field measurements of N_2O, CH_4, CO_2 fluxes from cultivated land. Atmos Environ 43:1888–1896

Zheng X, Xie B, Liu C, Zhou Z, Yao Z, Wang Y, Wang Y, Yang L, Zhu J, Huang Y, Butterbach-Bahl K (2008a) Quantifying Net Ecosystem Carbon Dioxide Exchange (NEE) of a short-plant cropland with intermittent chamber measurements. Glob Biogeochem Cycles 22, GB3031

Zheng X, Mei B, Wang Y, Xie B, Wang Y, Dong H, Xu H, Chen G, Cai Z, Yue J, Gu J, Su F, Zou J, Zhu J (2008b) Quantification of N2O fluxes from soil-plant systems may be biased by the applied gas-chromatography methodology. Plant Soil 311:211–234

Chapter 5
A Comparison of Methodologies for Measuring Methane Emissions from Ruminants

John P. Goopy, C. Chang, and Nigel Tomkins

Abstract Accurate measurement techniques are needed for determining greenhouse gas (GHG) emissions in order to improve GHG accounting estimates to IPCC Tiers 2 and 3 and enable the generation of carbon credits. Methane emissions from agriculture must be well defined, especially for ruminant production systems where national livestock inventories are generated. This review compares measurement techniques for determining methane production at different scales, ranging from in vitro studies to individual animal or herd measurements. Feed intake is a key driver of enteric methane production (EMP) and measurement of EMP in smallholder production systems face many challenges, including marked heterogeneity in systems and feed base, as well as strong seasonality in feed supply and quality in many areas of sub-Saharan Africa.

In vitro gas production studies provide a starting point for research into mitigation strategies, which can be further examined in respiration chambers or ventilated hood systems. For making measurements under natural grazing conditions, methods include the polytunnel, sulfur hexafluoride (SF_6), and open-path laser. Developing methodologies are briefly described: these include blood methane concentration, infrared thermography, pH, and redox balance measurements, methanogen population estimations, and indwelling rumen sensors.

J.P. Goopy (✉)
International Livestock Research Institute (ILRI),
Old Naivasha Rd., P.O. Box 30709, Nairobi, Kenya
e-mail: j.goopy@cgiar.org

C. Chang
Commonwealth Scientific and Industrial Research Organisation (CSIRO),
Townsville, QLD, Australia

N. Tomkins
Commonwealth Scientific and Industrial Research Organisation (CSIRO),
Livestock Industries, Townsville, QLD 4811, Australia

T.S. Rosenstock et al. (eds.), *Methods for Measuring Greenhouse Gas Balances and Evaluating Mitigation Options in Smallholder Agriculture*,
DOI 10.1007/978-3-319-29794-1_5

5.1 Introduction

Fermentation processes by rumen microbes result in the formation of reduced cofactors, which are regenerated by the synthesis of hydrogen (H_2) (Hungate 1966). Accumulation of excessive amounts of H_2 in the rumen negatively affects the fermentation rate and growth of some microbial consortia. Methanogens therefore reduce carbon dioxide (CO_2) to methane (CH_4) and water (H_2O) thereby capturing available hydrogen (McAllister et al. 1996). It is predicted that total CH_4 emissions from livestock in Africa will increase to 11.1 mt year^{-1} by 2030, an increase of 42 % over three decades (Herrero et al. 2008). Production increases and efficiencies in the livestock sector are seen as complementary outcomes if enteric methanogenesis can be reduced. While mitigation strategies are focused on manipulation of nutritional factors and rumen function, animal breeding programmes for selecting highly efficient animals that produce less enteric CH_4 might also be useful. Regardless of the mitigation strategy imposed, any reduction in enteric methane production (EMP) must be quantified and for this to be achieved, accurate baseline emissions data are essential.

This chapter reviews the existing and developing methodologies for gathering accurate data on ruminant methane production under a wide range of production systems. The principles of using predictive algorithms based on dietary, animal and management variables are considered here for modelling smallholder livestock emissions, but not in detail. Predictive models have been considered in detail elsewhere (Blaxter and Clapperton 1965; Kurihara et al. 1999; Ellis et al. 2007, 2008; Charmley et al. 2008; Yan et al. 2009). Major techniques are highlighted at different levels—in vitro, animal, herd and farm scale—and their advantages and disadvantages, including implementation in practice, are discussed. These methodologies can be used to support mitigation strategies or quantify total national livestock emissions.

5.2 Indirect Estimation

5.2.1 *In Vitro Incubation*

The amount of gas released from the fermentation process and the buffering of volatile fatty acids (VFAs) is related to the kinetics of fermentation of a known amount of feedstuff (Dijkstra et al. 2005). Several systems have been developed for measuring in vitro gas production, varying considerably in complexity and sophistication. Menke et al. (1979) describes a manual method using gastight syringes, which involves constant registering of the gas volume produced. More recently others have described a system using pressure transducers (Pell and Schofield 1993; Theodorou et al. 1994; Cone et al. 1996). Variants of this system are now available as proprietary systems (RF, ANKOM Technology®) using radiofrequency pressure sensor modules, which communicate with a computer interface and dedicated software to record gas pressure values.

The basic principle of the in vitro technique relies on the incubation of rumen inoculum with a feed substrate under an anaerobic environment in gastight culture bottles. Gas accumulates throughout the fermentation process and a cumulative volume is recorded. Gas volume curves can be generated over time. To estimate kinetic parameters of total gas production, gas production values are corrected for the amount of gas produced in a blank incubation and these values can be fitted with time using a nonlinear curve fitting procedure in GenStat (Payne et al. 2011) or other suitable software. Headspace gas samples are taken to analyze the gas compositions and determine actual CH_4 concentrations, typically by gas chromatography. A "quick and dirty" alternative is to introduce a strongly basic solution, such as NaOH into the vessel, which will cause the CO_2 to enter the solution. The remaining gas is assumed to be CH_4.

Gas is only one of the outputs of microbial fermentation, and the quality of the information derived can be improved by also considering substrate disappearance and production of VFAs (Blümmel et al. 2005).

5.2.2 *Estimation from Diet*

EMP can be estimated from intake and diet quality (digestibility). A number of algorithms can be used to do this, although estimates of emissions can vary by 35 % or more for a particular diet (Tomkins et al. 2011). Diet quality can be inferred from analysis of representative samples of the rations or pasture consumed, but where intake is not measured, estimation of EMP faces considerable challenges. Models which estimate intake based on diet quality or particular feed fractions assume ad libitum access, and in situations where animals are corralled without access to feed overnight, the validity of this assumption is likely violated (Jamieson and Hodgson 1979; Hendricksen and Minson 1980). In such a case, intake can be inferred from energy requirement (Live Weight (LW) + Energy for: LW flux; maintenance + lactation and pregnancy + locomotion) using published estimates (such as National Research Council) to convert physical values into energy values and so infer intake of the estimated diet. If this method is chosen, multiple measurements are required to capture changes in these parameters, as well as seasonal influences on feed availability and quality. Where possible, estimates made using this methodology should be validated by measurements in respiratory chambers.

5.3 Direct Measurement

5.3.1 *Open-Circuit Respiration Chambers*

Models to estimate national and global CH_4 emissions from sheep and cattle at farm level are mostly based on data of indirect calorimetric measurements (Johnson and Johnson 1995). Respiration chambers are used to measure CH_4 at an individual animal level. Their use is technically demanding, and only a few animals can be monitored at any one time (McGinn et al. 2008). However, these systems are capable of providing continuous and accurate data on air composition over an extended period of time.

Although the design of chambers varies, the basic principle remains the same. Sealed and environmentally controlled chambers are constructed to house test animals. All open-circuit chambers are characterized by an air inlet and exhaust, so animals breathe in a one-way stream of air passing through the chamber space. Air can be pulled through each chamber and, by running intake and exhaust fans at different speeds, negative pressure can be generated within the chamber. This is to ensure that air is not lost from the chamber (Turner and Thornton 1966). However, CH_4 can still be lost from chambers that are imperfectly sealed (down the concentration gradient), so gas recovery is an essential routine maintenance task. Thresholds for chamber temperature (<27 °C), relative humidity (<90 %), CO_2 concentration (<0.5 %), and ventilation rate (250–260 L min^{-1}) have been described (Pinares-Patiño et al. 2011), but may vary in practice. It is very important, however, to ensure that test animals remain in their thermoneutral zone while being measured, or intake is likely to be compromised. Some chambers may be fitted with air-conditioning units, which provide a degree of dehumidification and a ventilation system. This ensures that chambers can be maintained at constant temperature (Klein and Wright 2006) or at near-ambient temperature to capture normal diurnal variance (Tomkins et al. 2011). Choices about temperature are governed by technical resources and experimental objectives. Feed bins and automatic water systems may also be fitted with electronic scales and meters to monitor feed and water intake.

Change in O_2, CO_2, and CH_4 concentrations is measured by sampling incoming and outgoing air, using gas analyzers, infrared photoacoustic monitors, or gas chromatography systems (Klein and Wright 2006; Grainger et al. 2007; Goopy et al. 2014b). The other essential measurement is airflow, over a period of either 24 or 48 h. The accuracy and long-term stability of the measurements are dependent on the sensitivity of the gas analyzers used and the precision of their calibration. Chambers are directly calibrated by releasing a certain amount of standard gas of known concentration to estimate recovery values (Klein and Wright 2006). Measurement outcomes are also influenced by the environmental temperature, humidity, pressure, incoming air composition, and chamber volume. The larger the chamber, the less sensitive the measurements are to spatial fluctuations, as the response time is dependent on the size of the chamber and the ventilation rate (Brown et al. 1984). The calibration of the gas analyzers must be accurate and replicable for long-term use.

One constraint of this technique is that normal animal behavior and movement are restricted in the respiration chambers. Animals benefit from acclimatization in chambers prior to confinement and measurement, in order to minimize alterations in behavior, such as decreased feed intake (McGinn et al. 2009). However, there is clear evidence that this will happen in a small proportion of animals, regardless of training (Robinson et al. 2014) and this should be borne in mind when interpreting data. Using transparent construction material in chamber design allows animals to have visual contact with the other housed animals.

There are high costs associated with the construction and maintenance of open-circuit respiration chambers. The need for high performance and sensitive gas analyzers and flow meters must be considered in design and construction. Only a few animals can be used for measurements within chambers at any one time (Nay et al. 1994). Nevertheless, respiration chambers are suitable for studying the differences between treatments for mitigation strategies, and continue to be regarded as the "gold standard" for measuring individual emissions.

5.3.2 Ventilated Hood System

The ventilated hood system is a simplification of the whole animal respiration chamber, as it measures the gas exchange from the head only, rather than the whole body. Moreover, it is an improvement on face masks as used by Kempton et al. (1976), because gas measurements can be generated throughout the day and animals are able to access food and water.

Modern ventilated hood systems for methane measurements have been used in Japan, Thailand (Suzuki et al. 2007, 2008), USA (Place et al. 2011), Canada (Odongo et al. 2007) and Australia (Takahashi et al. 1999). Fernández et al. (2012) describes a mobile, open-circuit respiration system.

The ventilated hood system used by Suzuki et al. (2007, 2008) consists of a head cage, the digestion trial pen, gas sampling and analysis, behavior monitoring, and a data acquisition system. Similarly to whole animal chambers, it is equipped with a digestion pen for feed intake and excreta output measurements. An airtight head cage is located in front of the digestion pen and is provided with a loose fitting sleeve to position the animal's head. Head boxes are provided with blowers, to move the main air stream from the inlet to the exhaust. Flow meters correct the air volume for temperature, pressure, and humidity. Air filters remove moisture and particles from the gas samples, which are sent to the gas analyzers (Suzuki et al. 2007). The mobile system of Fernández et al. (2012) contains a mask or a head hood connected to an open-circuit respiration system, which is placed on a mobile cart.

The ventilated hood system is a suitable method under some circumstances, especially where open-circuit chambers are not viable. A critical limitation of the hood system is that extensive training is absolutely essential to allow the test animals to become accustomed to the hood apparatus. Thus while it can be used to assess potential of feeds, it is not suitable for screening large numbers of animals. A further consideration is that hoods capture only measurements of enteric methanogenesis and exclude the proportion emitted as flatus.

5.3.3 Polytunnel

Polytunnels are an alternative to respiration chambers, and operation and measurements are somewhat simpler. Methane emissions from individual or small groups of animals can be acquired under some degree of grazing. This allows test animals to express normal grazing behavior, including diet selection over the forages confined within the polytunnel space (Table 5.1). They have been used in the UK to measure

Table 5.1 Techniques for estimation of methane emission from livestock

Method	Description	Suitability	Cost	Accuracy and precision	Key references
Methods for indirect estimation					
These methods estimate CH_4 emissions without direct measurements on animals					
1. Lab-based (in vitro) incubation	Feed substrate is incubated in airtight bottles/bags to allow gas accumulation, and then gas samples analyzed for CH_4 concentrations	Can be used as a first approach to test potential feedstuffs and additives under controlled conditions	Less expensive and time consuming than respiration chambers	May not represent whole-animal (in vivo) emissions	Menke et al. (1979) (manual), Pell and Schofield (1993) (computerized)
2. Estimation from diet (models)	CH_4 is estimated from feed intake using models, usually developed from previous experimental data	Applicable in cases where measurements are not possible	Inexpensive to use once developed; eliminates need for CH_4 measurement	The assumptions and conditions that must be met for each equation limits their ability to accurately predict methane production	
		Requires estimates of feed intake, which can be challenging to obtain			
Methods for direct measurement of daily methane production (DMP)					
These methods monitor emissions continuously for extended periods and can be used to measure DMP					

3. Open circuit respiration chambers Cole et al. (2013)	Measures methane concentration within exhaled breath while the animal is in an enclosed chamber	Not suitable for examining effects of grazing management	Expensive to construct and maintain. Use is technically demanding	Provides most accurate and precise measurements of emissions, including CH_4 from ruminal and hindgut fermentations	Pinares and Waghorn (2014), Pinares-Patiño et al. (2011)
		Restricts normal animal behavior and movement; may decrease feed intake			
		Only a few animals can be used for measurement at any one time			
4. Ventilated hood Place et al. (2011)	An airtight box is placed to surround the animal's head. Gas exchange is measured only from the head rather than the whole body	Can be used to assess emissions from different feeds	Lower cost than whole-animal chamber	Does not measure hindgut CH_4	Fernández et al. (2012), Place et al. (2011), Suzuki et al. (2007)
		Restricts normal animal behavior and movement; not suitable for grazing systems	Requires training to allow the test animals to become accustomed to the hood apparatus		

(continued)

Table 5.1 (continued)

5. SF_6 tracer technique  Mottram	A small permeation tube containing SF_6 is placed in the cow's rumen, and SF_6 and CH_4 concentrations are measured near the mouth and nostrils of the cow	Allows the animal to move about freely; suitable for grazing systems	Lower cost, but higher level of equipment failure and more labor-intensive than respiration chambers	Less precise than respiration chambers	Johnson et al. (1994), Deighton et al. (2014), Berndt et al. (2014)
		Can be used to measure large numbers of individual animals	Animal must be trained to wear a halter and collection yoke	Does not measure hindgut CH_4	
		The challenge is that SF_6 itself is a GHG			
6. Polyethylene 	A large tunnel made of heavy-duty polyethylene fitted with end walls and large diameter ports. The concentrations of air between the incoming and outgoing air are continuously monitored	Suitable for measuring CH_4 emissions under semi-normal grazing conditions		With frequent calibration, provides high methane recovery rate, similar to respiration chambers	Lockyer and Jarvis (1995)
		Can be used for individual or small group of animals	Operation simpler than respiration chambers	There is difficulty in controlling the tunnel's temperature and humidity	
		As with SF_6 chambers, does not capture feed intake, so not suited for evaluating differences between imposed experimental treatments	Portable		

7. Open-path laser Mingenew Irwin	Lasers and wireless sensor networks send beams of light across paddocks containing grazing animals. The reflected light is analyzed for greenhouse gas concentrations	Measures CH_4 emissions from herds of animals and facilitates whole-farm measurements across a number of pastures	Expensive. Requires sensitive instrumentation to analyze CH_4 concentration and capture micrometeorological data	Accuracy is highly dependent on environmental factors and the location of test animals. Data must be carefully screened	Tomkins et al. (2011), Denmead (2008), Loh et al. (2008), Gao et al. (2010)
		Emissions cannot be attributed to a single source	Equipment requires continuous monitoring. Technically demanding		
Methods for short-term measurements					
These methods measure emissions over a short period, which can be used to estimate DMP or relative methane emissions					
8. Greenfeed® Emission Monitoring Apparatus Sutton (2014)	Patented device that measures and records short-term (3–6 min) CH_4 emissions from individual cattle repeatedly over 24 h by attracting animals to the unit using a "bait" of pelleted concentrate	Suitable for comparing effects of feeds or supplements	Patented device; must be purchased from supplier, C-lock Inc. (Rapid City, South Dakota, USA)	Provides comparable estimates to respiratory chamber and SF_6 techniques	Zimmerman and Zimmerman (2012), Hammond et al. (2013)
		Requires the use of a feed "attractant" to lure the animal to the facility, which alters results		Does not measure hindgut CH_4	

(continued)

Table 5.1 (continued)

9. Portable accumulation chambers John Goopy	Clear polycarbonate box in which the animal is placed for approximately 1 h; methane production is measured by the increase in concentration that occurs during that time	Designed to measure large numbers of animals for genetic screening of relative methane production Tested with sheep	Similar in cost to open-circuit respiration chambers, but with much shorter measurement time	Comparability with respiration chambers unclear. Further investigation is required before committing significant resources to this method	Goopy et al. (2011), Robinson et al. (2015)

CH_4 emissions from ruminants under semi-normal grazing conditions. Murray et al. (2001) reports CH_4 emissions from sheep grazing two ryegrass pastures and a clover–perennial ryegrass mixed pasture using this methodology. Essentially polytunnels consist of one large inflatable or tent type tunnel made of heavy duty polyethylene fitted with end walls and large diameter ports. Air is drawn through the internal space at speeds of up to 1 $m^3\ s^{-1}$ (Lockyer and Jarvis 1995). In general they are used where emissions from fresh forages are of interest because animals can be allowed to graze a confined area of known quality and quantity. When the available forage is depleted the tunnel is moved to a new patch.

Air flow rate can be measured at the same interval as the CH_4 or can be continuously sampled at the exhaust port (Lockyer 1997). Micropumps may be used to pass the exhausted air to a dedicated gas analyzer or a gas chromatograph (GC) (Murray et al. 2001). Data from all sensors can be sent to a data logger, which captures flow rate, humidity, and temperature within the tunnel, and gas production from the livestock. Samples of the incoming and exhaust air can be taken as frequently as necessary, depending on the accuracy required. The samples can be either taken manually or by an automatic sampling and injection system.

The polytunnel system requires frequent calibration to assure a good recovery rate, which is performed using the same principle as the chamber technique. Methane measurements can be collected over extended periods of time. Fluctuations occur due to changes in animal behavior, position relative to the exhaust port, internal temperature, relative humidity, and grazing pattern of the animal: eating, ruminating, or resting (Lockyer and Jarvis 1995; Lockyer and Champion 2001). The polytunnel is suitable for measuring CH_4 emissions under semi-normal grazing conditions. It has been reported that the polytunnel method gives 15 % lower readings of CH_4 concentration compared to the respiration chamber method, suggesting that animals actually consume less in the polytunnel. This requires further investigation. Recovery rate is high in both systems: 95.5–97.9 % in polytunnels, compared to 89.2–96.7 % in chambers (Murray et al. 1999). With an automated system, measurements can be performed with high repeatability. The system is portable and can be used on a number of pastures or browse shrubs, though again its utility is limited by the inability to capture feed intake.

5.3.4 Sulfur Hexafluoride Tracer Technique

The sulfur hexafluoride (SF_6) technique provides a direct measurement of the CH_4 emission of individual animals. This technique can be performed under normal grazing conditions, but can also be employed under more controlled conditions where intake is measured and/or regulated.

The SF_6 principle relies on the insertion of a permeation tube with a predetermined release ratio of SF_6 into the rumen, administered by mouth (Johnson et al. 1994). Air from around the animal's muzzle and mouth is drawn continuously into an evacuated canister connected to a halter fitted with a capillary tube around the neck. Johnson et al. (1994) provide a detailed description of the methodology.

The duration of collection of each sample is regulated by altering the length and/or diameter of the capillary tube (Johnson et al. 1994). Several modifications have since been reported with specific applications (Goopy and Hegarty 2004; Grainger et al. 2007; Ramirez-Restrepo et al. 2010). Most recently Deighton et al. (2014) has described the use of an orifice plate flow restrictor which considerably reduces the error associated with sample collection and should be considered in preference to the traditional capillary tube flow restrictors. At completion of sample collection the canisters are pressurized with N_2 prior to compositional analysis by gas chromatography. Enteric CH_4 production is estimated by multiplying the CH_4/SF_6 ratio by the known permeation tube release rate, corrected for actual duration of sample collection, and background CH_4 concentration (Williams et al. 2011), which is determined by sampling upwind ambient air concentration. Williams et al. (2011) emphasized the importance of correct measurement and reporting of the background concentrations, especially when the method is applied indoors. CH_4 is lighter (16 g mol^{-1}) than SF_6 (146 g mol^{-1}) and will therefore disperse and accumulate differently depending on ventilation, location of the animals, and other building characteristics.

This method enables gas concentrations in exhaled air of individual animals to be sampled and takes into account the dilution factor related to air or head movement. The high within- and between-animal variation is a significant limitation of this method. Grainger et al. (2007) reported variation within animals between days of 6.1 % and a variation among animals of 19.7 %. Pinares-Patiño et al. (2011) monitored sheep in respiration chambers simultaneously with the SF_6 technique. They reported higher within (×2.5) and between (×2.9) animal variance compared to the chamber technique, combined with a lower recovery rate (0.8 ± 0.15 with SF_6 versus 0.9 ± 0.10 with chambers). These sources of variation need to be taken into account in order to determine the number of repeated measures necessary to ensure accurate results. Moate et al. (2015) describes the use of Michaelis–Menten kinetics to better predict the discharge rate of capsules, which should reduce error associated with estimating discharge rates. It should also prolong the useful life of experimental subjects through the improved predictability of discharge rates over much longer intervals.

The SF_6 technique allows animals to move and graze normally on test pastures. This makes the method suitable for examining the effect of grazing management on CH_4 emissions (Pinares-Patiño et al. 2007) but it does so at a cost. The SF_6 method is less precise, less physically robust (high equipment failures), and more labor-intensive than respiration chamber measures.

5.3.5 *Open-Path Laser*

The use of open-path lasers combined with a micrometeorological dispersion method can now be used to measure enteric methane emissions from herds of animals. It therefore facilitates whole-farm methane measurements across a number of pastures.

The open-path laser method for whole-farm methane measurements is already in use in Canada (McGinn 2006; Flesch et al. 2005, 2007), Australia (Loh et al. 2008; McGinn et al. 2008; Denmead 2008; Tomkins et al. 2011), New Zealand (Laubach and Kelliher 2005) and China (Gao et al. 2010). Methane concentration measurements are performed using one or more tuneable infrared diode lasers mounted on a programmable and motorized scanning unit (Tomkins et al. 2011). The tuneable infrared diode laser beams to a retro reflector along a direct path, which reflects the beam back to a detector. The intensity of the received light is an indicator of the CH_4 concentration (ppm) along the path. In an optimal situation there should be at least one path for each predominant wind direction: one path upwind (background CH_4) and multiple paths downwind (CH_4 emission) of the herd. This method assumes that the herd acts as a surface source or, when individual animals can be fitted with GPS collars, individual animals are treated as point sources.

Regardless of application, the CH_4 concentration is calculated as the ratio of the external absorption to internal reference-cell absorption of the infrared laser beam as it travels along the path (Flesch et al. 2004, 2005). Methane concentration and environmental indicators such as atmospheric temperature, pressure, and wind direction and speed are continually measured and recorded using a weather station (Loh et al. 2008, 2009). Data—including GPS coordinates of the paddock or individual animals from a number of averaging time periods—can be merged using statistical software. After integrating, WindTrax software (Thunder Beach Scientific, Nanaimo, Canada) uses a backward Langrangian Stochastic (bLS) model to simulate CH_4 emissions (g day^{-1} per animal), by computing the line average CH_4 concentrations with atmospheric dispersion conditions.

The data integrity of the open-path laser method is highly dependent on environmental factors and the location of test animals. Flesch et al. (2007) described several criteria to determine data integrity using the open-path laser method. These criteria are based on wind turbulence statistics, laser light intensity, R^2 of a linear regression between received and reference waveforms, surface roughness, atmospheric stability, and the source location (surface or point source). Invalid data can be generated as a result of misalignment of the laser, unfavourable wind directions, surface roughness or periods in which the atmospheric conditions (rain, fog, heat waves, etc.) are unsuitable for applying the model (Freibauer 2000; Laubach and Kelliher 2005; Loh et al. 2008). To optimize the positioning of the equipment, these meteorological and physical aspects of the experimental site must be taken into account (Flesch et al. 2007; Loh et al. 2008, 2009). Moreover, the measurement area is restricted by the length of the laser paths when using a surface source approach. It is important to define the herd location, as uneven distribution of the herd results in miscalculations of the CH_4 concentration. Tomkins et al. (2011), comparing open-circuit respiration chambers with the open-path laser technique, reported estimated CH_4 emissions using the bLS dispersion model of 29.7 ± 3.70 g kg^{-1} dry matter intake (DMI), compared to 30.1 ± 2.19 g kg^{-1} DMI measured using open-circuit respiration chambers.

The open-path laser method does not interfere with the normal grazing behavior of the cattle and is noninvasive. Spatial variability is taken into account in these measurements, as the method can simulate gas fluxes over a large grazing area.

Moreover, the tuneable diode laser is highly sensitive and has a fast response to changes in CH_4 concentration, with detection limits at a scale of parts per trillion (McGinn et al. 2006). The labor intensity is low, although the equipment requires continuous monitoring. This method is expensive, which reflects not only the requirement for sensitive and rapid-response instruments to analyze CH_4 concentration, but also the requirement to capture micrometeorology data. Diurnal variations due to grazing and rumination pattern, pasture composition, and individual variation need to be considered in planning experimental protocols to prevent over- or undercalculation of the total emission. Furthermore, DMI determination is not very accurate as this is based on predictive models using the relationship between LW and LW gain, following assumption of the ARC (1980).

5.4 Short-Term Measurement

While most assessments of enteric methane emissions are focused on daily methane production (DMP), or the derivative, daily methane yield (MY), there is an increasing impetus to estimate the emissions of large numbers of animals in their productive environment. This is driven both by the demand for data to establish genetic parameters for DMP and to verify mitigation strategies or GHG inventories. This area is discussed only briefly here, as there is currently limited scope for the application of these technologies in sub-Saharan Africa. The area has been ably reviewed by Hegarty (2013).

5.4.1 Greenfeed® Emission Monitoring Apparatus

Greenfeed® is a patented device (Zimmerman and Zimmerman 2012) that measures and records short-term (3–6 min) CH_4 emissions from individual cattle repeatedly over 24 h by attracting animals to the unit using a "bait" of pelleted concentrate. By being available 24 h day^{-1} potential sampling bias is reduced and the technique has been shown to provide comparable estimates to those produced both by respiratory chamber and SF_6 techniques (Hammond et al. 2013). However, a significant limitation of the technique is the requirement to supply an "attractant" to lure the animal to use the facility, consisting of up to 1 kg of concentrate pellets per day. This will certainly affect DMP and may also alter VFA profiles or the overall digestibility of the diet. Attempts to use energy neutral attractants, such as water have proven equivocal (J Velazco, personal communication).

5.4.2 Portable Accumulation Chambers

Portable accumulation chambers (PAC) consist of a clear polycarbonate box of approximately 0.8 m^3 volume, open at the bottom and sealed by achieving close contact with flexible rubber matting. Methane production is measured by the increase in concentration that occurs while an animal is in the chamber for approximately 1 h. PACs were designed to screen large numbers of sheep, variously to identify potentially low and high emitting individuals and to develop genetic parameter estimates in sheep populations. This technique initially showed close agreement with respiratory chamber measurements (Goopy et al. 2009, 2011). Subsequent investigations demonstrated such measurements to be moderately repeatable in the field and to have potential for genetic screening of animals (Goopy et al. 2015). Longer-term comparisons of PAC measurements and respiratory chamber data, however, suggest that these two methods may be measuring quite different traits and further investigation is required before committing significant resources to PAC measurements (Robinson et al. 2015).

5.4.3 *Application of CH_4:CO_2 Ratio*

Madsen et al. (2010) proposed using the ratio of CH_4:CO_2 in exhaled breath to assess EMP in ruminants. This method requires knowledge about the intake, energy content, and heat increment of the ration consumed. Haque et al. (2014) applied this method, using a fixed heat increment factor. Hellwing et al. (2013) regressed open-circuit chamber measurements of DMP in cattle against estimates calculated using CH_4:CO_2 ratios and found them to be only moderately correlated ($R^2=0.4$), which suggest this method is unsuitable for precision measurements.

5.4.4 *Spot Sampling with Lasers*

Spot measurements of methane in the air around cattle's mouths have been made using laser devices to provide short-term estimates of enteric methane flux (Chagunda et al. 2009; Garnsworthy et al. 2012). These estimates are then scaled up to represent DMP – requiring an impressive number of assumptions to be met to satisfy such scaling. Chagunda and Yan (2011) have claimed correlations of 0.7 between laser and respiratory chamber measurements, but this claim is based on the laser apparatus measuring methane concentrations in the outflow of the chambers, rather than from the animals themselves.

5.5 Emerging and Future Technologies

5.5.1 *Blood Methane Concentration*

This methodology relies on enteric methane being absorbed across the rumen wall, transported in the blood stream to the pulmonary artery and respired by the lungs. The jugular (vein) gas turnover rate of enteric SF_6 (introduced by an intraruminal bolus) and CH_4 has been used to determine the respired concentrations and solubility of these gases (Ramirez-Restrepo et al. 2010). The solubility coefficients and CH_4 concentrations are determined by gas chromatography, comparing the peak area of the sampled gases with standards. Variances in CH_4 and SF_6 blood concentrations may be related to the methodology, or may occur because these gases are not equally reabsorbed. This requires further investigation. Sampling can be logistically challenging and labor-intensive and it is important to recognize that this method provides little more than a "snapshot" of methane concentration at the time of sampling.

5.5.2 *Infrared Thermography*

Montanholi et al. (2008) have examined the use of infrared thermography as an indicator for heat and methane production in dairy cattle. No direct relationship was reported, however, between temperature in any specific part of the body and methane production.

5.5.3 *Intraruminal Telemetry*

The use of a rumen bolus to measure methane in the liquid phase is logistically possible and small changes ($<50\ \mu mol\ L^{-1}$) in CH_4 concentrations could be detectable (Gibbs 2008). Low pH and redox potential have been correlated with decreased CH_4 concentrations, and a pH and redox sensor have been developed to suit a rumen bolus by eCow Electronic Cow Management at the University of Exeter, UK (www.ecow.co.uk). This technology is still in its exploratory stages but the application of a rumen bolus to measure CH_4 in the rumen headspace has been patented (McSweeney, personal communication.) and could theoretically provide accurate CH_4 concentration estimates for large numbers of free grazing animals.

5.5.4 *Quantitative Molecular Biology*

Gibbs (2008) examined the correlation between the numbers of methanogens and CH_4 production in short time intervals. Results from real-time polymerase chain reaction (PCR) suggest that increased CH_4 production is related to increased methanogen metabolic activity rather than increased population size.

5.6 Summary

EMP is a complex trait, involving animal physiology and behavior, plant factors, and animal management. Although there are many techniques available to estimate EMP, all have limitations. The appropriateness of a technique is strongly influenced by its intended purpose and the degree of precision required. It is important to recognize that while more sophisticated in vitro techniques can provide robust information about the fermentative, and hence, methanogenic potential of feeds, they do not truly represent in vivo fermentation, nor do they account for feed intake, and will be of limited predictive use for animals grazing heterogeneous pastures. If intake is unknown it will diminish the utility of established models, especially when assumptions regarding *ad libitum* intake are violated. Lasers, infrared, and SF_6 techniques can all be used to measure EMP of animals at pasture. However, all are technically fastidious and in situations where intake is unknown, cannot be used to determine emissions intensity. Respiration chambers, while requiring significant capital to construct and technical skill to operate, provide precise and accurate measurements of EMP on known feed intake. Whilst there are justified criticisms surrounding reproducibility of EMP at pasture and evidence of changed feeding behavior in some cases, respiration chambers remain the most accurate method of assessing EMP in individual animals.

References

ARC (1980) The nutrient requirements of ruminant livestock. Australian Research Council, CAB International, Wallingford

Berndt A, Boland TM, Deighton MH, Gere JI, Grainger C, Hegarty RS, Iwaasa AD, Koolaard JP, Lasse KR, Luo D, Martin RJ, Martin C, Moate PJ, Molano G, Pinares-Patiño C, Ribaux BE, Swainson NM, Waghorn GC, Williams SRO (2014) Guidelines for use of sulphur hexafluoride (SF 6) tracer technique to measure enteric methane emissions from ruminants. In: Lambert MG (ed). New Zealand Agricultural Greenhouse Gas Research Centre, New Zealand. www.globalresearchalliance.org/app/uploads/2012/03/SF6-Guidelines-all-chapters-web.pdf. Accessed 15 March 2015

Blaxter KL, Clapperton JL (1965) Prediction of the amount of methane produced by ruminants. Br J Nutr 19:511–522

Blümmel M, Givens DI, Moss AR (2005) Comparison of methane produced by straw fed sheep in open-circuit respiration with methane predicted by fermentation characteristics measured by an in vitro gas procedure. Anim Feed Sci Technol 123–124(part 1):379–390

Brown D, Cole TJ, Dauncey MJ, Marrs RW, Murgatroyd PR (1984) Analysis of gaseous exchange in open-circuit indirect calorimetry. Med Biol Eng Comput 22:333–338

Chagunda MGG, Yan T (2011) Do methane measurements from a laser detector and an indirect open-circuit respiration calorimetric chamber agree sufficiently closely? Anim Feed Sci Technol 165:8–14

Chagunda MGG, Ross D, Roberts DJ (2009) On the use of a laser methane detector in dairy cows. Comput Electron Agric 68:157–160

Charmley E, Stephens ML, Kennedy PM (2008) Predicting livestock productivity and methane emissions in northern Australia: development of a bio-economic modelling approach. Aust J Exp Agric 48:109–113

Cole NA, Hales KE, Todd, RW, Casey K, MacDonald JC (2013) Effects of corn processing method and dietary inclusion of wet distillers grains with solubles (WDGS) on enteric methane emissions of finishing cattle. Waste to worth: spreading science and solutions, Denver, 1–5 April 2013. http://www.extension.org/pages/67580/effects-of-corn-processing-method-and-dietary-inclusion-of-wet-distillers-grains-with-solubles-wdgs-#.VhfI5LRViko. Accessed 9 Oct 2015

Cone JW, van Gelder AH, Visscher GJW, Oudshoorn L (1996) Influence of rumen fluid substrate concentration on fermentation kinetics measured with a fully automated time related gas production apparatus. Anim Feed Sci Technol 61:113–128

Deighton MH, Williams SRO, Hannah MC, Eckard RJ, Boland TM, Wales WJ, Moate PJ (2014) A modified sulphur hexafluoride tracer technique enables accurate determination of enteric methane emissions from ruminants. Anim Feed Sci Technol 197:47–63

Denmead OT (2008) Approaches to measuring fluxes of methane and nitrous oxide between landscapes and the atmosphere. Plant Soil 309:5–24

Dijkstra J, Kebreab E, Bannink A, France J, López S (2005) Application of the gas production technique to feed evaluation systems for ruminants. Anim Feed Sci Technol 123–124:561–578

Ellis JL, Kebreab E, Odongo NE, McBride BW, Okine EK, France J (2007) Prediction of methane production from dairy and beef cattle. J Dairy Sci 90:3456–3467

Ellis JL, Dijkstra J, Kebreab E, Bannink A, Odongo NE, McBride BW, France J (2008) Modelling animal systems paper: aspects of rumen microbiology central to mechanistic modelling of methane production in cattle. J Agric Sci 146:213–233

Fernández C, López MC, Lachica M (2012) Description and function of a mobile open-circuit respirometry system to measure gas exchange in small ruminants. Anim Feed Sci Technol 172:242–246

Flesch TK, Wilson JD, Harper LA, Crenna BP, Sharpe RR (2004) Deducing ground-to-air emissions from observed trace gas concentrations: a field trial. J Appl Meteorol 43:487–502

Flesch TK, Wilson JD, Harper LA, Crenna BP (2005) Estimating gas emission from a farm using an inverse dispersion technique. Atmos Environ 39:4863–4874

Flesch TK, Wilson JD, Harper LA, Todd RW, Cole NA (2007) Determining ammonia emissions from a cattle feedlot with an inverse dispersion technique. Agr Forest Meteorol 144:139–155

Freibauer A (2000) New approach to an inventory of N_2O and CH_4 emissions from agriculture in Western Europe. Kluwer Academic, Dordrecht

Gao Z, Desjardins RL, Flesch TK (2010) Assessment of the uncertainty of using an inverse-dispersion technique to measure methane emissions from animals in a barn and in a small pen. Atmos Environ 44:3128–3134

Garnsworthy PC, Craigon J, Hernandez-Medrano JH, Saunders N (2012) On-farm methane measurements during milking correlate with total methane production by individual dairy cows. J Dairy Sci 95:3166–3180

Gibbs J (2008) Novel methane assessment in ruminants. Final report for project code: CC MAF POL_2008-37. Ministry of Agriculture and Forestry, Government of New Zealand, Wellington. http://maxa.maf.govt.nz/climatechange/slm/grants/research/2007-08/2008-37-summary.htm. Accessed 15 March 2015

Goopy J, Hegarty R (2004) Repeatability of methane production in cattle fed concentrate and forage diets. J Anim Feed Sci 13:75–78

Goopy J, Hegarty R, Robinson D (2009) Two hour chamber measurement provides a useful estimate of daily methane production in sheep. In: Ruminant physiology: digestion, metabolism and effects of nutrition on reproduction and welfare. Proceedings of the XIth international symposium on ruminant physiology, Clermont-Ferrand, France, 6–9 Sept 2009. Wageningen Academic, The Netherlands, p 190

Goopy JP, Woodgate R, Donaldson A, Robinson DL, Hegarty RS (2011) Validation of a short-term methane measurement using portable static chambers to estimate daily methane production in sheep. Anim Feed Sci Technol 166–167:219–226

Goopy JP, Donaldson A, Hegarty R, Vercoe PE, Haynes F, Barnett M, Oddy VH (2014) Low-methane yield sheep have smaller rumens and shorter rumen retention time. Br J Nutr 111:578–585

Goopy J, Robinson D, Woodgate R, Donaldson A, Oddy H, Vercoe P, Hegarty R (2015) Estimates of repeatability and heritability of methane production in sheep using portable accumulation chambers. Anim Prod Sci. http://dx.doi.org/10.1071/AN13370

Grainger C, Clarke T, McGinn SM, Auldist MJ, Beauchemin KA, Hannah MC, Waghorn GC, Clark H, Eckard RJ (2007) Methane emissions from dairy cows measured using the sulfur hexafluoride (SF_6) tracer and chamber techniques. J Dairy Sci 90:2755–2766

Hammond K, Humphries D, Crompton L, Kirton P, Green C, Reynolds C (2013) Methane emissions from growing dairy heifers estimated using an automated head chamber (GreenFeed) compared to respiration chambers or SF_6 techniques. Adv Anim Biosci 4:391

Haque MN, Cornou C, Madsen J (2014) Estimation of methane emission using the CO_2 method from dairy cows fed concentrate with different carbohydrate compositions in automatic milking system. Livest Sci 164:57–66

Hegarty RS (2013) Applicability of short-term emission measurements for on-farm quantification of enteric methane. Animal 7:401–408

Hellwing A, Lund P, Madsen J, Weisberg MR (2013) Comparison of enteric methane production from the CH_4/CO_2 ratio and measured in respiration chambers. Adv Anim Biosci 4:557

Hendricksen R, Minson DJ (1980) The feed intake and grazing behaviour of cattle grazing a crop of Lablab purpureus cv. Rongai. J Agric Sci 95:547–554

Herrero M, Thornton PK, Kruska R, Reid RS (2008) Systems dynamics and the spatial distribution of methane emissions from African domestic ruminants to 2030. Agr Ecosyst Environ 126:122–137

Hungate RE (1966) The rumen and its microbes. Academic, New York

Jamieson WS, Hodgson J (1979) The effects of variation in sward characteristics upon the ingestive behaviour and herbage intake of calves and lambs under a continuous stocking management. Grass Forage Sci 34:273–282

Johnson KA, Johnson DE (1995) Methane emissions from cattle. J Anim Sci 73:2483–2492

Johnson K, Huyler M, Westberg H, Lamb B, Zimmerman P (1994) Measurement of methane emissions from ruminant livestock using a SF_6 tracer technique. Environ Sci Technol 28:359–362

Kempton T, Murray R, Leng R (1976) Methane production and digestibility measures in the grey kangaroo and sheep. Aust J Biol Sci 29:209–214

Klein L, Wright ADG (2006) Construction and operation of open-circuit methane chambers for small ruminants. Aust J Exp Agric 46:1257–1262

Kurihara M, Magner T, Hunter RA, McCrabb GJ (1999) Methane production and energy partition of cattle in the tropics. Br J Nutr 81:227–234

Laubach J, Kelliher JM (2005) Methane emissions from dairy cows: comparing open-path laser measurements to profile-based techniques. Agr Forest Meteorol 135:340–345

Lockyer DR (1997) Methane emission from grazing sheep and calves. Agr Ecosyst Environ 66:11–18

Lockyer DR, Champion RA (2001) Methane production in sheep in relation to temporal changes in grazing behavior. Agr Ecosyst Environ 86:237–246

Lockyer DR, Jarvis SC (1995) The measurement of methane losses from grazing animals. Environ Pollut 90:383–390

Loh Z, Chen D, Bai M, Naylor T, Griffith D, Hill J, Denmead T, McGinn S, Edis R (2008) Measurement of greenhouse gas emissions from Australian feedlot beef production using open-path spectroscopy and atmospheric dispersion modelling. Aust J Exp Agric 48:244–247

Loh Z, Leuning R, Zegelin S, Etheridge D, Bai M, Naylor T, Griffith D (2009) Testing Lagrangian atmospheric dispersion modeling to monitor CO_2 and CH_4 leakage from geosequestration. Atmos Environ 43:2602–2611

Madsen J, Bjerg BS, Hvelplund T, Weisbjerg MR, Lund P (2010) Methane and carbon dioxide ratio in excreted air for quantification of the methane production from ruminants. Livest Sci 129:223–227

McAllister TA, Okine EK, Mathison GW, Cheng KJ (1996) Dietary, environmental and microbiological aspects of methane production in ruminants. Can J Anim Sci 76:231–243

McGinn SM (2006) Measuring greenhouse gas emissions from point sources in agriculture. Can J Soil Sci 86:355–371

McGinn SM, Flesch TK, Harper LA, Beauchemin KA (2006) An approach for measuring methane emissions from whole farms. J Environ Qual 35:14–20

McGinn SM, Chen D, Loh Z, Hill J, Beauchemin KA, Denmead OT (2008) Methane emissions from feedlot cattle in Australia and Canada. Aust J Exp Agric 48:183–185

McGinn SM, Beauchemin KA, Flesch TK, Coates T (2009) Performance of a dispersion model to estimate methane loss from cattle in pens. J Environ Qual 38:1796–1802

Menke KH, Raab L, Salewski A, Steingass H, Fritz D, Schneider W (1979) The estimation of the digestibility and metabolizable energy content of ruminant feeding stuffs from the gas production when they are incubated with rumen liquor in vitro. J Agric Sci 93:217–222

Moate PJ, Deighton MH, Ribaux BE, Hannah MC, Wales WJ, Williams SRO (2015) Michaelis–Menten kinetics predict the rate of SF_6 release from permeation tubes used to estimate methane emissions from ruminants. Anim Feed Sci Technol 200:47–56

Montanholi YR, Odongo NE, Swanson KC, Schenkel FS, McBride BW, Miller SP (2008) Application of infrared thermography as an indicator of heat and methane production and its use in the study of skin temperature in response to physiological events in dairy cattle (Bos taurus). J Therm Biol 33:468–475

Mottram T (no date) Monitoring the rumen. eCow. http://research.ecow.co.uk/publications/rumen-monitoring. Accessed 9 Oct 2015

Murray PJ, Moss A, Lockyer DR, Jarvis SC (1999) A comparison of systems for measuring methane emissions from sheep. J Agric Sci 133:439–444

Murray PJ, Gill E, Balsdon SL, Jarvis SC (2001) A comparison of methane emissions from sheep grazing pastures with different management intensities. Nutr Cycl Agroecosyst 60:93–97

Nay SM, Mattson KG, Bormann BT (1994) Biases of chamber methods for measuring soil CO_2 efflux demonstrated with a laboratory apparatus. Ecology 75:2460–2463

Odongo NE, Alzahal O, Las JE, Kramer A, Kerrigan B, Kebreab E, France J, McBride BW (2007) Development of a mobile open-circuit ventilated hood system for measuring real-time gaseous emissions in cattle. CABI Publishing, Wallingford

Payne RW, Murray DA, Harding SA (2011) An introduction to the GenStat Command Language, 14th edn. VSN International, Hemel Hempstead

Pell AN, Schofield P (1993) Computerized monitoring of gas production to measure forage digestion in vitro. J Dairy Sci 76:1063–1073

Pinares C, Waghorn G (2014) Technical manual on respiration chamber designs. New Zealand Ministry of Agriculture and Forestry, Wellington. http://www.globalresearchalliance.org/app/uploads/2012/03/GRA-MAN-Facility-BestPract-2012-FINAL.pdf. Accessed 15 March 2015

Pinares-Patiño CS, Hour PD, Jouany JP, Martin C (2007) Effects of stocking rate on methane and carbon dioxide emissions from grazing cattle. Agr Ecosyst Environ 121:30–46

Pinares-Patiño CS, Lassey KR, Martin RJ, Molano G, Fernandez M, MacLean S, Sandoval E, Luo D, Clark H (2011) Assessment of the sulphur hexafluoride (SF_6) tracer technique using respiration chambers for estimation of methane emissions from sheep. Anim Feed Sci Technol 166:201–209

Place SE, Pan Y, Zhao Y, Mitloehner FM (2011) Construction and operation of a ventilated hood system for measuring greenhouse gas and volatile organic compound emissions from cattle. Animal 1:433–446

Ramirez-Restrepo CA, Barr TN, Marriner A, López-Villalobos N, McWilliam EL, Lassey KR, Clark H (2010) Effects of grazing willow fodder blocks upon methane production and blood composition in young sheep. Anim Feed Sci Technol 155:33–43

Robinson D, Goopy J, Donaldson A, Woodgate R, Oddy V, Hegarty R (2014) Sire and liveweight affect feed intake and methane emissions of sheep confined in respiration chambers. Animal 8:1935–1944

Robinson DL, Goopy JP, Hegarty RS, Oddy VH (2015) Comparison of repeated measurements of CH4 production in sheep over 5 years and a range of measurement protocols. J Anim Sci 93:4637–4650

Sutton M (2014) Life's a gas for unitrial cattle. Stock J. http://www.stockjournal.com.au/news/agriculture/cattle/beef/lifes-a-gas-for-uni-trial-cattle/2692193.aspx. Accessed 9 Oct 2015

Suzuki T, McCrabb GJ, Nishida T, Indramanee S, Kurihara M (2007) Construction and operation of ventilated hood-type respiration calorimeters for in vivo measurement of methane production and energy partition in ruminants. Springer, Dordrecht

Suzuki T, Phaowphaisal I, Pholson P, Narmsilee R, Indramanee S, Nitipot T, Haokaur A, Sommar K, Khotprom N, Panichpol V, Nishida T (2008) In vivo nutritive value of pangola grass (Digitaria eriantha) hay by a novel indirect calorimeter with a ventilated hood in Thailand. Jarq-Jpn Agric Res Q 42:123–129

Takahashi J, Chaudhry AS, Beneke RG, Young BA (1999) An open-circuit hood system for gaseous exchange measurements in small ruminants. Small Rumin Res 32:31–36

Theodorou MK, Williams BA, Dhanoa MS, McAllan AB, France J (1994) A simple gas production method using a pressure transducer to determine the fermentation kinetics of ruminant feeds. Anim Feed Sci Technol 48:185–197

Tomkins NW, McGinn SM, Turner DA, Charmley E (2011) Comparison of open-circuit respiration chambers with a micrometeorological method for determining methane emissions from beef cattle grazing a tropical pasture. Anim Feed Sci Technol 166–167:240–247

Turner HG, Thornton RF (1966) A respiration chamber for cattle. Aust Soc Anim Prod 6:413–419

Williams SRO, Moate PJ, Hannah MC, Ribaux BE, Wales WJ, Eckard RJ (2011) Background matters with the SF_6 tracer method for estimating enteric methane emissions from dairy cows: a critical evaluation of the SF_6 procedure. Anim Feed Sci Technol 170:265–276

Yan T, Porter MG, Mayne CS (2009) Prediction of methane emission from beef cattle using data measured in indirect open-circuit respiration calorimeters. Animal 3:1455–1462

Zimmerman PR, Zimmerman RS (2012) Method and system for monitoring and reducing ruminant methane production. United States Patent number US20090288606A

Chapter 6
Quantifying Tree Biomass Carbon Stocks and Fluxes in Agricultural Landscapes

Shem Kuyah, Cheikh Mbow, Gudeta W. Sileshi, Meine van Noordwijk, Katherine L. Tully, and Todd S. Rosenstock

Abstract This chapter presents methods to quantify carbon stocks and carbon stock changes in biomass of trees in agricultural landscapes. Specifically it assesses approaches for their applicability to smallholder farms and other tree enterprises in agricultural landscapes. Measurement techniques are evaluated across three criteria: accuracy, cost, and scale. We then recommend techniques appropriate for users looking to quantify carbon in tree biomass at the whole-farm and landscape scales. A basic understanding of the carbon cycle and the concepts of biomass assessment is assumed.

6.1 Introduction

Trees and woody biomass play an important role in the global carbon cycle. Forest biomass accounts for over 45 % of terrestrial carbon stocks, with approximately 70 % and 30 % contained within the above and belowground biomass, respectively (Cairns et al. 1997; Mokany et al. 2006). Not all trees exist inside forests, however.

S. Kuyah (✉)
World Agroforestry Centre (ICRAF), UN Avenue-Gigiri, PO Box 30677-00100, Nairobi, Kenya

Jomo Kenyatta University of Agriculture and Technology (JKUAT), Nairobi, Kenya
e-mail: s.kuyah@cgiar.org

C. Mbow • T.S. Rosenstock
World Agroforestry Centre (ICRAF), UN Avenue-Gigiri, PO Box 30677-00100, Nairobi, Kenya

G.W. Sileshi
Freelance Consultant, 5600 Lukanga Road, Kalundu, Lusaka, Zambia

M. van Noordwijk
World Agroforestry Centre (ICRAF), Bogor, Indonesia

K.L. Tully
University of Maryland, College Park, MD, USA

T.S. Rosenstock et al. (eds.), *Methods for Measuring Greenhouse Gas Balances and Evaluating Mitigation Options in Smallholder Agriculture*,
DOI 10.1007/978-3-319-29794-1_6

Table 6.1 Typical precision for various quantification uses

End user	Potential uses	Typical precision
National governments	Reporting to the IPCC	Variable
	Development of National Appropriate Mitigation Actions	
Markets	Carbon trading between governments and businesses	±10 to 20 %
Development organizations	Promotion of low emission agricultural development	Undefined

Trees feature prominently in agricultural landscapes globally. Almost half of all agricultural lands maintain at least 10 % tree cover (Zomer et al. 2014) (Table 6.1). Despite widespread distribution, tree outside forests (TOF) are an often neglected carbon pool and little information is available on carbon stocks in these systems or their carbon sequestration potential (de Foresta et al. 2013; Hairiah et al. 2011).

The ubiquity and use of trees in agricultural landscapes is significant for smallholder farmers' livelihoods and modifying local climate (van Noordwijk et al. 2014), but it also contributes to global climate change mitigation (Nair et al. 2009, 2010). Even when planted at low densities, the aggregate carbon accumulation in trees can help fight climate change because of the large spatial extent covered (Verchot et al. 2007; Zomer et al. 2014). Such trees are estimated to accumulate 3–15 Mg ha^{-1} $year^{-1}$ in aboveground biomass alone (Nair et al. 2010), a non-trivial amount when compared to other carbon sinks such as soil. Simultaneously, trees diversify diets, reduce soil erosion, and expand market opportunities for smallholder farmers (Van Noordwijk et al. 2011). Thus, trees in agricultural landscapes offer opportunities to mitigate climate change and improve smallholder livelihoods (Kumar and Nair 2011). The synergy between climate adaptation and mitigation through trees in agricultural lands is now receiving explicit attention (Duguma et al. 2014).

Despite the significant advances in assessment methods, quantifying carbon stocks and fluxes at different spatial scale is still challenging. Although National Forest Inventories (NFIs) are supposed to provide such guidelines, they are well developed only in the Northern hemisphere. Most NFIs also do not include trees outside forests (TOF) and until recently TOF have been poorly defined (de Foresta et al. 2013; Baffetta et al. 2011). Hence sampling designs that can be consistently applied to both forests and TOF are lacking while ideally national biomass estimates should include carbon estimates of both forests and TOF. Most NFIs (except Sweden and Canada) do not include explicit TOF categories (de Foresta et al. 2013).

The dearth of consistent methodology and a new interest to integrate trees in farming systems in global biomass assessments (de Foresta et al. 2013) is catalyzing efforts to generate data on biomass and carbon specific for trees on farmland. This, however, comes with the challenge to rapidly develop and standardize methods for biomass assessment, obstacles in the forestry community has been grappling with for decades. Forest-based methodologies can be adapted for some applications. However, TOF present unique issues. To begin with, tree stands in agricultural landscapes typically show irregular shapes when compared to those in more dense forest stands. The geometry of tree stands on farmland is particularly plastic, sensitive to local environmental conditions (Harja et al. 2012), and human management (Dossa et al. 2007; Frank and Eduardo 2003). Tree management (pruning, coppicing, lopping, etc.) may violate assumptions of the available allometries, which were developed based on physiological relationships (e.g., mass and diameter at breast height (DBH)) observed in forests and plantations (Kuyah et al. 2012a). The impact of local edaphic conditions on tree growth combined with the diversity of uses and agroecological conditions complicates the construction of a coherent database to represent carbon and biomass estimation equations (BEMs) for farmland trees. The consequence is a scarcity of data and a fragmented understanding of the role trees on farms may play in climate and development discussions.

With more attention paid to farm forestry, agroforestry, and expansion of the agricultural frontier in many countries, quantification of biomass in trees in agricultural landscapes is receiving greater attention. There is a growing interest in the assessments of carbon stocks and sequestration for carbon monitoring and reporting needs, but also as a way to evaluate agricultural interventions (Thangata and Hildebrand 2012). In the following sections, we discuss general considerations of measurement accuracy, cost, and scale when quantifying and discuss the two predominant quantification approaches for biomass and carbon in trees on farms.

6.2 Accuracy, Scale, and Cost

Accurate estimates of changes in C stocks are required and uncertainties should be reduced as much as is practical (IPCC 2003). Yet, uncertainty depends strongly on scale and the costs of high accuracy plus high spatial resolution must be weighed against the benefits of farmer incentive schemes that need such information, as opposed to cheaper solutions that meet accuracy targets by spatial aggregation, e.g., to a 1 km^2 scale (Lusiana et al. 2014). Methodological limitations and random as well as systematic errors associated with quantification of biomass of trees on farms guarantee uncertainties in estimates. A large degree of uncertainty exists in estimations of C stocks and fluxes at the local, regional, and global scale. Some of the uncertainty results from the lack of consensus on definitions, inconsistencies in methods, and assumptions leading to widely differing results even among similar studies (Sileshi 2014). These variations are mainly a result of lack of a common framework for sampling. Uncertainty in C estimation should be addressed to establish the reliability of estimates and provide a basis of confidence for decision-making, particularly where comparisons (e.g., with baseline results) are involved. Identified uncertainties can be quantified through statistical methods such as error propagation (Chave et al. 2004). Uncertainties in biomass quantification result from six primary sources in the quantification process: (1) the level of detail in the method used, (2) the complexities of the systems and landscapes being modeled, (3) sampling error, (4) measurement error, (5) model errors, and (6) the inconsistency in

Table 6.2 Comparison of approaches and techniques in terms of scale, cost, and accuracy

Approach	Scale	Cost	Accuracy	Uncertainty
Destructive sampling	Limited to small area	Expensive	Most accurate	
Allometry	Allows upscaling	Cheap once equations are developed	Relatively accurate	
Dendrochronology	High resolution at tree level	Cheap once the lab equipment exists	Very accurate if individual rings are easy to read	Missing rings, wood anatomy, wedging, etc.
Remote sensing	Variable (high to low resolution)	High-resolution data are still very expensive	Relatively accurate depending on the indices of method used	For low resolution there is blended information that reduce farm-level assessment

estimating and reporting biomass components (Chave et al. 2004; IPCC 2003). Available biomass and carbon estimates for trees on farms vary considerably and associated measures of uncertainty in the estimates (e.g., standard errors and confidence intervals) are often not reported.

There is a potential mismatch between the scale at which measurements are made and the scale at which information is required for policy and programmatic development. Different methodologies allow quantification of carbon stocks at various spatial and temporal scales, ranging from plot to landscape scale and shorter and longer time horizons. Here again, the method used depends on the available funds and accuracy required. Field sampling methods destructive (i.e., harvesting trees, drying, and weighing biomass) or non-destructive (i.e., use of BEMs) are affordable and applicable for only a limited number of sites (Table 6.2). Remote sensing is practical and effective for mapping aboveground biomass in expansive remote areas, e.g., at regional scale.

The cost of carbon quantification depends on the method chosen, a choice that is determined by the scale of measurement and desired level of accuracy. The methods presented here vary in their degree of robustness, allowing for trade-offs between accuracy, cost, and practical viability for smallholder systems (Table 6.2). The key is to determine information that can be obtained at relatively low cost but still produces estimates within an acceptable level of accuracy. Destructive measurements are known to be costly in terms of resources, effort, and time, and are not permitted for rare or protected species. Modeling with BEMs is therefore an expedient way of estimating carbon both from field inventories or remote sensing. Obtaining field inventories is expensive, slow, and impractical in large areas. Ground-based measurements of tree diameters are therefore often combined with predictive models to estimate carbon stocks in small areas that can be upscaled. The costs on field inventories and analytical methods are greatly influenced by the sampling design used and the minimum number of measurement required for a particular method. For both modeling with BEMs and remote sensing, costs can be greatly reduced and efficiencies of labor and time achieved by adopting multipurpose sampling sites or procedures. For example, the sites could be designed to take measurements for carbon quantification, and also provide data for biodiversity analyses or assessment of vegetation and soil properties. An example is the Land Health Surveillance Framework, designed to cost-effectively enable measurement and monitoring of carbon in a given landscape over years (Vågen et al. 2010). Regarding the models, simple power-law models with DBH alone are less expensive to develop and use compared to parameter-rich models. This is because DBH measurements can be easily obtained at low cost compared to specialized equipment required for height or crown area measurements. Remote sensing can greatly reduce the time and cost of collecting data over large areas, particularly for highly variable, widely spaced, and hard-to-access areas (Wulder et al. 2008). However, remote sensing approaches such as airplane-mounted LiDAR instruments are still too costly and technically demanding. And while remote-sensing instruments can estimate proxies that can also be converted into biomass using statistical models; additional expenses will be incurred on field data for calibration/validation, which are also prone to errors. This is because there is no remote-sensing instrument that can presently measure tree carbon stocks directly (Gibbs et al. 2007).

6.3 Quantification of Five Carbon Pools of Representative Plots

Tree biomass can be estimated using direct (destructive) or indirect (non-destructive) approaches (Pearson et al. (2005) or GOFC-GOLD (2011) for methods, models, and parameters widely used). Direct methods require felling of trees and weighing the component parts. Destructive sampling provides the best data for building BEMs, generating inventory for estimating biomass, and providing requisite information for validating indirectly estimated biomass (Brown 1997; Gibbs et al. 2007). By contrast, indirect methods (e.g., BEMs and remote sensing) use readily measurable proxies, such as DBH, crown area, or vegetation indices that are then converted into biomass based on statistical relationships established by destructive sampling (Brown 2002; Bar Massada et al. 2006). Unfortunately, most algorithms and regressions relating remotely sensed data to biomass increase precision, not accuracy. Therefore, it is important to make ground measurements to increase the accuracy of BEMs and remotely sensed data.

Cost considerations require that estimates of carbon stocks and stock changes on farms and landscapes be based on representative samples from land uses and covers and measurement of proxy variables rather than quantifying biomass on every farm or pixel and destructive sampling of trees, respectively. Indirect measures and statistical models only approximate biomass with a precision subject to the representativeness of the models to local conditions. That latter consideration is particularly salient for smallholder situations in tropical developing countries. Models have largely been constructed on data not collected in the tropics and little in Africa (Hofstad 2005; Henry et al. 2011) and even fewer data and BEMs are available for trees on farms. Applying equations to data with size range beyond the one that was used in building the equations can lead to high levels of bias and poor estimates of biomass. Biomass—and carbon—estimates by indirect methods will therefore always be inaccurate. Qualitatively, at least, the direct linkage between tree architecture, as modified by farm management, and fractal branching models that generate allometric equations suggests ways to make adjustments where major branches or parts of the crown are missing from trees (Hairiah et al. 2011; MacFarlane et al. 2014).

The cost and time of destructive measurement make it impractical for most uses. Therefore, this discussion focuses on indirect quantification methods. Indirect quantification of four IPCC identified biomass carbon pools (aboveground biomass, belowground biomass, deadwood, and litter) involves a series of steps (1) stratification/identification of the target areas, (2) measurement of proxies for biomass, (3) calculating biomass/carbon (4) scaling to whole-farms and landscapes (Fig. 6.1). This highlights the need to recognize two aspects to the uncertainty of carbon estimation: the first aspect is plot level—how good are measurements of biomass in the field? Do they account for belowground biomass, dead biomass, soil carbon, hollow trees, and smaller trees e.g., those <10 cm diameter? How good are we at converting wood volumes into total aboveground biomass? The second aspect of uncertainty is converting plot-level measurements across space, either through modeling or with satellite data.

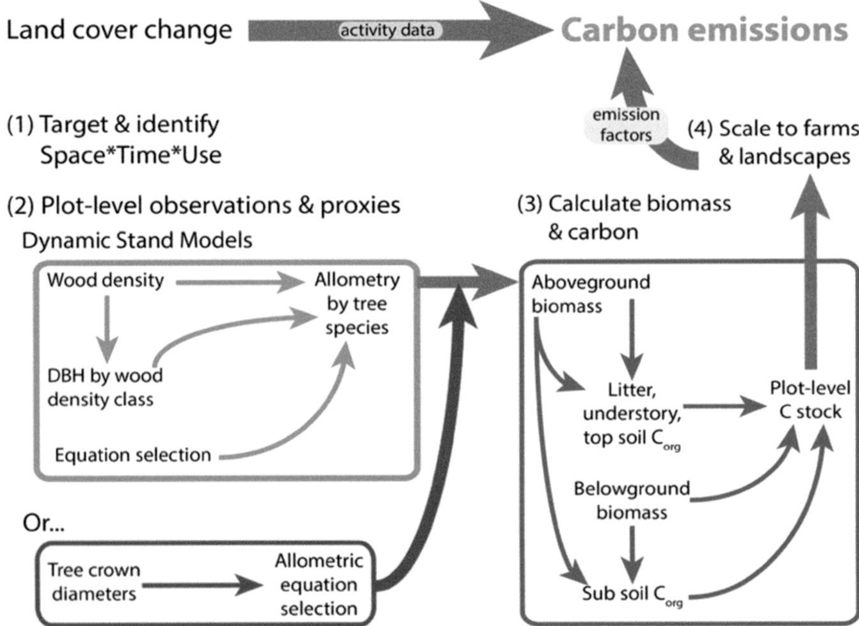

Fig. 6.1 Mixed-method approach to five-pool carbon estimates for farms and landscapes

6.3.1 Selecting Plots

Quantification at farm and landscape scales requires extrapolation from data gathered from relatively small plots to larger areas. Extrapolation is necessary because it is prohibitively expensive to measure every tree on every farm or throughout the landscape. With stratification, we aim to quantify the biomass and carbon at a few representative locations and then use data on the frequency of their occurrence to calculate total biomass at larger spatial extents. It is therefore critically important that the sample is representative of the larger area and farm/landscape features of interest and an estimate of the frequency of occurrence of the feature of interest is possible (Brown 1997). A stratified random sampling approach can be employed to guide sample selection ranging from remote sensing to household surveys. For building BEMs, a randomized pre-sample of trees can be generated from an inventory with respect to a stratified diameter class and trees for destructive sampling chosen through a blind selection without tree species association. For inventories, stratification by topographic features, management influence, and age classes are likely to produce more homogenous strata from which sample units could be selected. Age is essential particularly where lifecycle analysis is involved. In rotational plantations this is easy to implement, but in many land use systems derived from natural vegetation by selective retention of trees (e.g., shea or baobab trees in many savanna systems), regeneration pattern need to inform the sample selection. In systems with "internal regeneration," similar to natural forest with a gap renewal cycle, the age of the most frequent tree diameter class can be used to reconstruct a time-averaged carbon stock at the land use system level (Hairiah et al. 2011). We refer you to Chap. 2 of this manual and the references therein to determine an appropriate method for stratifying the sample. The remainder of this discussion assumes the availability of representative plots and knowledge of the relative distribution of different features or land use classes in the geographic space of interest.

6.3.2 Measurements of Proxies for Tree Biomass

Tree biomass is estimated from ground-based inventory data, remote sensing, or a combination of the two. Researchers and project developers tend to rely on BEMs, which calculate tree biomass based on easily measured dimensions based on the idea that standard relationships occur such as the diameter to mass or height to mass (West 2009) or root-to-shoot (Cairns et al. 1997; Mokany et al. 2006). Because of the variations in tree characteristics among ecological conditions, particularly in agricultural landscapes, and the need to account for biomass in all plant parts, it is ideal to use locally developed equations or develop BEMs at a local scale (Henry et al. 2011). Where local BEMs are not available, there are two other options. First, volume equation and inventory data arising from commercial interest valuing the stock of wood resources in forests may be available in many developing countries (Hofstad 2005; Henry et al. 2011). However, this approach provides data primarily on merchantable wood, leaving out components such as branches, twigs, and leaves, yet in some species these components constitute a significant amount, about 3 %, of the total aboveground biomass (Kuyah et al. 2013). The second option is to use the pantropical models (e.g., Chave et al. 2005). However, these are broadly derived, based on a large dataset and stratified by region or climatic conditions. The definition of climatic regimes is not intuitive and direct application of these models could give biased estimates if applied across the board, particularly in agricultural landscapes where trees face multiple stresses (Kuyah et al. 2012a; Sileshi 2014).

BEMs require the measurement of tree dimensions such as DBH, basal area, height, or crown dimensions. Presuming measurements are conducted with care, accurate biomass estimates are best obtained by measurements of each parameter. However, certain measurements (e.g., height) are difficult to obtain accurately in the field by non-destructive methods and hence including this parameter in models may introduce error into the biomass estimates, by a mean of 16 % (Hunter et al. 2013). Furthermore, complete datasets are in many cases not necessary to provide a reasonable estimate of biomass because inclusion of all parameters only moderately increases the accuracy of the total estimate. For example, inclusion of DBH alone provided an estimate within 1.5 % of the actual biomass measured in an agricultural landscape of Western Kenya (Kuyah and Rosenstock in review), which agrees with most studies (Cole and Ewel 2006; Basuki et al. 2009; Bastien-Henri et al. 2010). Given the complexities and potential errors in measuring other parameters (i.e., difficult terrain or dense foliage when measuring height), the need for specialized tools (e.g., hypsometer or clinometer for height), or destructive measurements (e.g., wood density), the use of DBH alone appears cost-effective and robust for most purposes (Sileshi 2014).

At landscape scales, ground-based inventories are typically too resource-intensive to complete. Instead, crown area—which can be measured by remote sensing—is increasingly being tested for estimating aboveground biomass (Wulder et al. 2008; Rasmussen et al. 2011; Fig. 6.2). Two issues complicate widespread application of remote sensing and crown areas. First, crown area is not as strongly correlated with biomass as DBH. This may be particularly important for trees on farms that show irregular growth patterns due to variable environmental conditions (e.g., near red/far red light interception, availability of soil nutrients) or management by farmers (e.g., limb collection for firewood). For example, (Kuyah et al. 2012b) show crown area measurements alone grossly misrepresent standing stocks of carbon, by about 20 % relative to diameter estimates. It is therefore important to calibrate remotely sensed crown area estimates with field measured DBH to improve the accuracy of measurements. Second, remote sensing of crown areas for trees outside of forests requires high-resolution imagery to differentiate small features such as individual trees on farms. Typically, Quickbird images with sub-m resolution are best suited for this task but cost ~15 USD per km. Without sufficient

Fig. 6.2 Delineation of TOF crowns by remote sensing using sub-meter resolution Quickbird imagery (Gumbricht unpublished)

resolution, it is not possible to identify trees and may lead to underestimation of biomass. Unfortunately, the price of the satellite imagery increases in parallel with the resolution and the specialized skills necessary to process the imagery limits many applications of this technique outside of the research arena at this time. Despite the challenges, crown area allometry is likely the most promising approach to transform our ability to capture information on aboveground biomass stocks, potentially for relatively low total costs in the future (Gibbs et al. 2007; Wulder et al. 2008).

Field measurements and remote sensing generate estimates of aboveground biomass. Though most of the carbon in trees is contained in aboveground biomass, a significant fraction can be found in the four other major carbon pools: belowground biomass, litter, deadwood, and soils. Soil carbon is discussed in Chap. 7 (Saiz and Albrech this volume) and thus we restrict this brief discussion to the other three pools. For almost all applications, belowground biomass will be estimated by allometric relationships based on DBH or prescribed root-to-shoot ratios. We are quite skeptical of the accuracy of general root-to-shoot ratios for estimation of belowground biomass as the growth patterns are sensitive to water availability and may range from 10:1 in moist conditions versus 4:1 in arid conditions (IPCC 2003). Recent destructive experiments suggest that DBH may be a better predictor than root-to-shoot ratio for trees on farms but again require inventories to establish DBH. Global studies show that belowground biomass (BG) is isometrically related to aboveground biomass (AG) (Hui et al. 2014; Cheng and Niklas 2007); i.e., BG = a(AG). If one can correctly estimate '*a*', we believe estimating BG from AG using allometric method may be better than using shoot-to-root ratios.

Consideration of litter and deadwood deserve unique attention for trees on farm. Litter might be assumed to be in equilibrium with growth and thus ignored in biomass estimation especially on farmland. Deadwood might also be treated in the same way given most will be collected for firewood or in slash and burn agriculture, fire will consume most of it. A case can be made that the relative limited size of these pools justifies such treatment for most cases, especially when considering decadal timescales. In cases when litter and deadwood need to be estimated, measurements using small nested plots or an independent sampling design will be required. For litter, the information collected is total mass per unit area but for dead wood, depending on the size, one can measure total mass or estimate volume that can be used for mass calculation if wood density is known (Pearson et al. 2005, 2007).

6.3.3 Calculating C Stocks and Fluxes

Until now, we have been discussing the quantification of biomass stocks in a small plot area. Oftentimes, however, researchers and project developers are more interested in the change in carbon, accumulation or loss, with various practices or land use change. So here we consider methods to quantify rates of change in woody biomass.

Time-Averaged Carbon Stock for Different Land Uses

Carbon stocks in trees generally accumulate slowly over time. Often it is therefore most appropriate to analyze the changes over multiple years or decadal time scales. On longer time scales it is possible to analyze the average change (per annum or a given time interval) for the lifecycle of the land use or farming system (see Fig. 6.3, for example). Stock change accounting assesses the magnitude of change carbon stored between two or more ecosystems that share a reference state. This approach is desirable because it allows a researcher to substitute space for time, overcoming the challenges of returning to measure the same location/land use/trees twice. Researchers locate farming systems existing in the landscape that have already been transformed from other land use systems. Carbon stocks calculated from the different systems can then be compared to provide a relative estimate of changes over time. Characteristically, the changes are standardized to changes per year. This approach assumes that carbon stock changes results from land use change/management and changes in carbon stocks are linear over the time period examined. This latter assumption negates the temporal dynamics of carbon. Yet, time averaged carbon stock presents a snapshot picture about the relative annual flux and cumulative impacts.

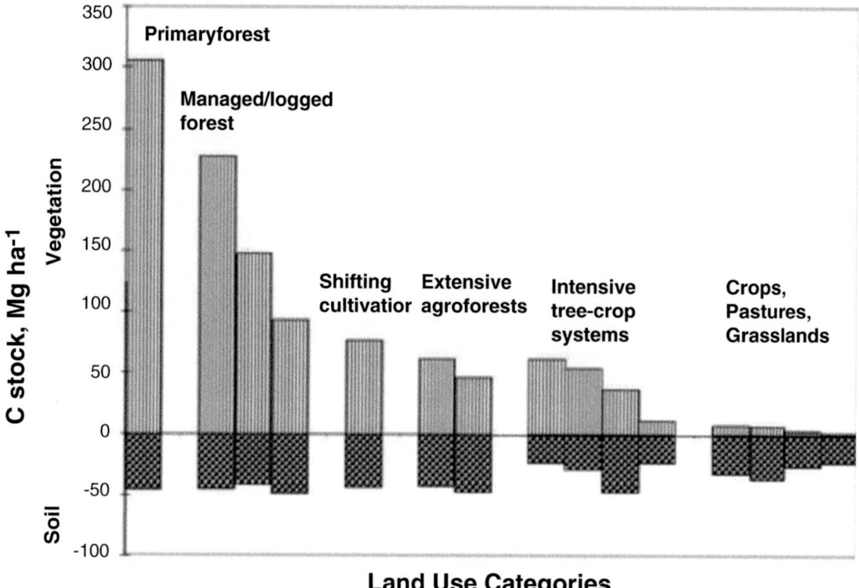

Fig. 6.3 Time-averaged aboveground carbon and total soil carbon (0–20 cm). *Source*: Hairiah et al. (2011)

Annual Changes: Growth Rates, Dendrochronology, Repeated Measurements

Though rarely quantified, examining annual changes in biomass carbon in trees on farm is important when calculating whole-farm GHG balances, that is, when calculating the global warming potential or global warming intensity of the system. Unfortunately, the growth rates of tropical tree species are only known for a small sample of commercially viable timber species and the remaining knowledge gap greatly limits the ability to map or model carbon stock changes. There are typically few options to gain information about annual stock changes in the absence of published growth rates: repeated measurements, biomass expansion factors (BEFs), and dendrochronology.

Repeated measurement of the same tree species is an option to create information on growth rates or annual changes in carbon stocks. Repeated measurements must be cautious to return precisely to the same tree/stand and the same measurement of the tree. Because repeated measurement relies on exact locations to document what can sometimes be small changes, this method is sensitive to observational and measurement errors as well as anomalies in growth patterns on the tree selected. Furthermore, repeated measurements can typically only be performed on a limited number of trees. Thus again, tree selection, to account for heterogeneity and minimize sampling artifacts, is critical. Though not without uncertainty, repeated measurement do provide a non-destructive approach to quantify short-term changes in carbon stocks.

BEFs are another approach of using exiting stand volume data from previous forest studies to assess carbon density. BEF bundles two aspects, a conversion of volume to mass and an inclusion of ignored trees foliage, small branches not accounted in commercial volume assessment. The BEF is a conversion factor that calculates biomass based on traditional commercial volume data (Brown 1997).

The use of dendrochronology is an emerging field of application of tree ring for biomass assessment for individual tree growth. The method is based on the formation of annual rings in many tropical trees in areas with one distinct dry season. Often, this seasonality induces cambial dormancy of trees, particularly if these belong to deciduous species (Brienen and Zuidema 2005). Annual tree rings provide growth information for the entire life of trees and their analysis has become more popular in tropical forest regions over the past decades (Soliz-Gamboa et al. 2010). It is demonstrated that tree-ring studies is a powerful tool to develop high-resolution and exactly dated proxies for biomass accumulation over time in individual trees (Mbow et al. 2013). In addition to annual increment of biomass, tree-ring analysis helps characterize climate–growth relationship between tree growth and rainfall in certain periods of the year and how this translates into tree productivity information that is central to carbon sequestration assessment (Mbow et al. 2013). Basically the use of such method implies the application of allometric models on diameter over bark on individual rings measured during the tree lifetime (Gebrekirstos et al. 2008). Important information can be collected using tree ring: (1) growth rate—average annual diameter increment-of-individual species to reconstruct long-term growth of trees and estimate productivity of trees; (2) age–diameter relationships which are required in carbon projections; (3) limiting factors of tree growth such as long time drought or severe fires.

6.3.4 Scaling to Whole-Farms and Landscapes

The final step involves aggregating the data on carbon stocks or stock changes into whole-farm and landscape-scale estimates. The precise scaling methods applied somewhat depend on the types of data collected and the equations used. However, scaling plot measurements will generally proceed in the following steps:

1. Land use/cover transition matrix in proportion for each zone by spatial analysis
2. Frequency of each zone by spatial analysis
3. Total area of the target area by spatial analysis (expressed in hectares)
4. Carbon stock of each system component calculated from the plot level measurements, allometric equations, and statistical analysis (expressed in Mg C ha^{-1})
5. Changes in the C stock for each transition by multiplying each cell in the matrix by the difference in the time-averaged C stock for each transition/conversion by the conversion factor (depending on plot size; expressed in Mg CO_2 equivalent ha^{-1})

6. Annual changes in C stock for each transition by dividing changes in C stock by the length of the study period (expressed in Mg CO_2 equivalent ha^{-1})
7. Total annual emission and total sequestration and net changes of C stock in the landscape (expressed in Mg CO_2 equivalent ha^{-1})

Because the principal scaling approach relies on similarity-based relationships (e.g., allometric equations) that are scale invariant, the same steps are equally relevant for whole-farms or landscapes, irrespective of the spatial extent. Furthermore, since the results are expressed in CO_2 equivalent ha^{-1} it is possible to integrate these measures with those from other GHG sources and sinks such as soil carbon or trace gas emissions from soils.

6.4 Additional Sources of Information

Because of the interest in forest inventories, there are countless sources of information available to help appropriately select and apply various techniques. Table 6.3 tabulates what we feel are the key sources of information, and links to specific protocols can be found on the website (http://www.samples.ccafs.cgiar.org/protocol/Biomass).

Table 6.3 Annotated key sources of information

Brown S (1997) Estimating Biomass and Biomass Change of Tropical Forests: a Primer. (FAO Forestry Paper—134). Food and Agriculture Organization of the United Nations (FAO), Rome, Italy
This report describes multiple methods for estimating biomass density, including one of the first comprehensive descriptions of methods for destructive biomass estimation. The report includes biomass estimates for different tropical countries based on forest type and climate. Supplementary tables report wood density for different tree species across tropical Asia, America, and Africa
West PW (2009) Tree and Forest Measurement. 2nd edition. Springer, Heidelberg, Germany
The primary audience for this book is undergraduate forestry students, practicing foresters, and landholders. As such, it introduces the techniques of tree and forest measurement with particular attention paid to non-destructive (allometric) approaches. This book provides a step-by-step description of how to measure trees as well as their component parts and then scale to the stand or population
One hundred years of tree-ring research in the tropics-a brief history and an outlook to future challenges. Dendrochronologia 20:217-231
This article describes the history of tree-ring analysis in the tropics. Tropical dendrochronology is hotly debated primarily because the consistent intra-annual temperatures of tropical systems do not produce the same tree-ring pattern we observe in temperate tree-rings. Worbes discusses the progress in and applications of tropical tree-ring research. One such application that we would like to highlight is the potential to use tree-rings to evaluate individual tree growth and thus track biomass accumulation through time

(continued)

Table 6.3 (continued)

Mbow C, Chhin S, Sambou B, Skole DL (2013) Potential of dendrochronology to assess annual rates of biomass productivity in savanna trees of West Africa. Dendrochronologia 31:41-51
This article describes the application of dendrochronology to assess biomass in individual savanna species in Southern Senegal. The materials and methods section provides a comprehensive description of the steps involved. Note that the destructive sampling was implemented in this study and may not be suitable for some situations
Gibbs HK, Brown S, Niles JO, Foley JA (2007) Monitoring and estimating tropical forest carbon stocks: making REDD a reality. Environ Res Lett 2:13
Using REDD (reductions in emissions from deforestation in developing countries) as a backdrop, this review paper discusses a range of methods available to estimate national-level forest carbon stocks using both ground-based and remotely sensed measurements of particular forest characteristics which can be converted into estimates of national carbon stocks using allometric relationships. This is the first article to report a complete set of national-level carbon stock estimates

References

Baffetta F, Corona P, Fattorini L (2011) Assessing the attributes of scattered trees outside the forest by a multi-phase sampling strategy. Forestry 84(3):315–325

Bar Massada A, Carmel Y, Tzur GE, Grünzweig JM, Yakir D (2006) Assessment of temporal changes in aboveground forest tree biomass using aerial photographs and allometric equations. Can J Forest Res 36(10):2585–2594

Bastien-Henri S, Park A, Ashton M, Messier C (2010) Biomass distribution among tropical tree species grown under differing regional climates. For Ecol Manage 260:403–410

Basuki TM, van Laake PE, Skidmore AK, Hussin YA (2009) Allometric equations for estimating the above-ground biomass in tropical lowland Dipterocarp forests. For Ecol Manage 257(8):1684–1694

Brienen RJ, Zuidema PA (2005) Relating tree growth to rainfall in Bolivian rain forests: a test for six species using tree ring analysis. Oecologia 146(1):1–12

Brown S (1997) Estimating biomass and biomass change of tropical forests: a primer. In: FAO Forestry paper no. 134. Food and Agriculture Organization of the United Nations (FAO), Rome

Brown S (2002) Measuring carbon in forests: current status and future challenges. Environ Pollut 116:363–372

Cairns MA, Brown S, Helmer EH, Baumgardner GA (1997) Root biomass allocation in the world's upland forests. Oecologia 111:1–11

Chave J, Condit R, Aguilar S, Hernandez A, Lao S, Perez R (2004) Error propagation and scaling for tropical forest biomass estimates. Philos Trans R Soc Lond B Biol Sci 359:409–420

Chave J, Andalo C, Brown S, Cairns MA, Chambers JQ, Eamus D, Fölster H, Fromard F, Higuchi N, Kira T, Lescure J-P, Nelson BW, Ogawa H, Puig H, Riéra B, Yamakura T (2005) Tree allometry and improved estimation of carbon stocks and balance in tropical forests. Oecologia 145(1):87–99

Cheng DL, Niklas KJ (2007) Above- and below-ground biomass relationships across 1534 forested communities. Ann Bot 99(1):95–102

Cole T, Ewel J (2006) Allometric equations for four valuable tropical tree species. For Ecol Manage 229:351–360

de Foresta H, Somarriba E, Temu A, Boulanger D, Feuilly H, Gauthier M (2013) Towards the assessment of trees outside forests. In: Resources assessment working paper 183. Food and Agriculture Organization of the United Nations (FAO), Rome

Dossa EL, Fernandes ECM, Reid WS, Ezui K (2007) Above- and belowground biomass, nutrient and carbon stocks contrasting an open-grown and a shaded coffee plantation. Agrofor Syst 72(2):103–115

Duguma L, Minang PA, van Noordwijk M (2014) Climate change mitigation and adaptation in the land use sector: from complementarity to synergy. Environ Manage 54(3):420–432

Frank B, Eduardo S (2003) Biomass dynamics of Erythrina lanceolata as influenced by shoot pruning intensity in Costa Rica. Agrofor Syst 57:19–28

Gebrekirstos A, Mitlöhner R, Teketay D, Worbes M (2008) Climate–growth relationships of the dominant tree species from semi-arid savanna woodland in Ethiopia. Trees 22(5):631–641

Gibbs HK, Brown S, Niles JO, Foley JA (2007) Monitoring and estimating tropical forest carbon stocks: making REDD a reality. Environ Res Lett 2:1748–9326

GOFC-GOLD (2011) A sourcebook of methods and procedures for monitoring and reporting anthropogenic greenhouse gas emissions and removals caused by deforestation, gains and losses of carbon stocks in forests remaining forests, and forestation. GOFC-GOLD Report version COP17-1. GOFC-GOLD Project Office, Natural Resources Canada, Alberta

Hairiah K, Dewi S, Agus F, Velarde S, Ekadinata A, Rahayu S, van Noordwijk M (2011) Measuring carbon stocks across land use systems: a manual. World Agroforestry Centre (ICRAF), SEA Regional Office, Bogor

Harja D, Vincent G, Mulia R, Van Noordwijk M (2012) Tree shape plasticity in relation to crown exposure. Trees 26:1275–1285

Henry M, Picard N, Trotta C, Manlay RJ, Valentini R, Bernoux M, Saint-André L (2011) Estimating tree biomass of Sub-Saharan African forests: a review of available allometric equations. Silva Fennica 45(3B):477–569

Hofstad O (2005) Review of biomass and volume functions for individual trees and shrubs in Southeast Africa. J Trop For Sci 17:151–162

Hui D, Wang J, Shen W, Le X, Ganter P, Ren H (2014) Near isometric biomass partitioning in forest ecosystems of China. PLoS One 9(1), e86550

Hunter MO, Keller M, Victoria D, Morton DC (2013) Tree height and tropical forest biomass estimation. Biogeosciences 10(12):8385–8399

IPCC (2003) Good practice guidance for land use, land-use change and forestry. Intergovernmental Panel on Climate Change (IPCC) National Greenhouse Gas Inventories Programme. Institute for Global Environmental Strategies (IGES), Kanagawa

Kumar BM, Nair PKR (2011) Carbon sequestration potential of agroforestry systems: opportunities and challenges. In: Advances in agroforestry, vol 8. Springer, New York

Kuyah S, Dietz J, Muthuria C, Jamnadassa R, Mwangi P, Coe R, Neufeldt H (2012a) Allometric equations for estimating biomass in agricultural landscapes: I. Aboveground biomass. Agric Ecosyst Environ 158:216–224

Kuyah S, Muthuri C, Jamnadass R, Mwangi P, Neufeldt H, Dietz J (2012b) Crown area allometries for estimation of aboveground tree biomass in agricultural landscapes of Western Kenya. Agrofor Syst 86(2):267–277

Kuyah S, Dietz J, Muthuri C, van Noordwijk M, Neufeldt H (2013) Allometry and partitioning of above- and below-ground biomass in farmed eucalyptus species dominant in Western Kenyan agricultural landscapes. Biomass Bioenergy 55:276–284

Lusiana B, van Noordwijk M, Johana F, Galudra G, Suyanto S, Cadisch G (2014) Implication of uncertainty and scale in carbon emission estimates on locally appropriate designs to reduce emissions from deforestation and degradation (REDD+). Mitig Adapt Strat Glob Chang 19(6):757–772

MacFarlane DW, Kuyah S, Mulia R, Dietz J, Muthuri C, van Noordwijk M (2014) Evaluating a non destructive method for calibrating tree biomass equations derived from tree branching architecture. Springer, Berlin

Mbow C, Chhin S, Sambou B, Skole DL (2013) Potential of dendrochronology to assess annual rates of biomass productivity in savanna trees of West Africa. Dendrochronologia 31:41–51

Mokany K, Raison RJ, Prokushkin AS (2006) Critical analysis of root shoot: ratios in terrestrial biomes. Glob Chang Biol 12:84–96

Nair PKR, Kumar BM, Nair VD (2009) Agroforestry as a strategy for carbon sequestration. J Plant Nutr Soil Sci 172:10–23

Nair PKR, Nair VD, Kumar BM, Showalter JM (2010) Carbon sequestration in agroforestry systems. In: Sparks DL, du Pont SH (eds) Advances in agronomy, vol 108, chap 5. Elsevier, Amsterdam

Pearson T, Walker S, Brown S (2005) Sourcebook for land use, land-use change and forestry projects. Winrock International. http://www.goldstandard.org/wp-content/uploads/2013/07/Winrock-BioCarbon_Fund_Sourcebook-compressed.pdf. Accessed 29 Jan 2015

Pearson TRH, Brown S, Birdsey RA (2007) Measurement guidelines for the sequestration of forest carbon. In: General technical report NRS-18. USDA Forest Service. http://www.nrs.fs.fed.us/pubs/gtr/gtr_nrs18.pdf. Accessed 29 Jan 2015

Rasmussen MO, Göttsche F-M, Diop D, Mbow C, Olesen F-S, Fensholt R, Sandholt I (2011) Tree survey and allometric models for tiger bush in northern Senegal and comparison with tree parameters derived from high resolution satellite data. Int J Appl Earth Obs Geoinf 13(4):517–527

Sileshi GW (2014) A critical review of forest biomass estimation models, common mistakes and corrective measures. For Ecol Manage 329:237–254

Soliz-Gamboa CC, Rozendaal DMA, Ceccantini G, Angyalossy V, Borg K, Zuidema PA (2010) Evaluating the annual nature of juvenile rings in Bolivian tropical rainforest trees. Trees 25(1):17–27

Thangata PH, Hildebrand PE (2012) Carbon stock and sequestration potential of agroforestry systems in smallholder agroecosystems of sub-Saharan Africa: mechanisms for 'reducing emissions from deforestation and forest degradation' (REDD+). Agric Ecosyst Environ 158:172–183

Vågen TG, Shepherd KD, Walsh MG, Winowiecki L, Desta LT, Tondoh JE (2010) AfSIS technical specifications: soil health surveillance. World Agroforestry Centre. http://worldagroforestry.org/sites/default/files/afsisSoilHealthTechSpecs_v1_smaller.pdf. Accessed 29 Jan 2015

Van Noordwijk M, Hoang MH, Neufeldt H, Öborn I, Yatich T (2011) How trees and people can co-adapt to climate change: reducing vulnerability in multifunctional landscapes. World Agroforestry Centre (ICRAF). http://www.worldagroforestry.org/sea/Publications/files/book/BK0149-11.PDF. Accessed 29 Jan 2015

van Noordwijk M, Bayala J, Hairiah K, Lusiana B, Muthuri C, Khasanah N, Mulia R (2014) Agroforestry solutions for buffering climate variability and adapting to change. In: Fuhrer J, Gregory PJ (eds) Climate change impact and adaptation in agricultural systems. CAB-International, Wallingford, pp 216–232

Verchot LV, Noordwijk M, Kandji S, Tomich T, Ong C, Albrecht A, Mackensen J, Bantilan C, Anupama KV, Palm C (2007) Climate change: linking adaptation and mitigation through agroforestry. Mitig Adapt Strat Glob Chang 12(5):901–918

West PW (2009) Tree and forest measurement, 2nd edn. Springer, Heidelberg

Wulder MA, White JC, Fournier RA, Luther JE, Magnussen S (2008) Spatially explicit large area biomass estimation: three approaches using forest inventory and remotely sensed imagery in a GIS. Sensors 8:529–560

Zomer RJ, Trabucco A, Coe R, Place F, van Noordwijk M, Xu JC (2014) Trees on farms: an update and reanalysis of agroforestry's global extent and socio-ecological characteristics. Working paper 179. World Agroforestry Centre (ICRAF) SEA Regional Program, Bogor

Chapter 7
Methods for Smallholder Quantification of Soil Carbon Stocks and Stock Changes

Gustavo Saiz and Alain Albrecht

Abstract Smallholder agricultural systems in tropical and subtropical regions may have significantly contributed to greenhouse gas (GHG) emissions over the past number of decades. As a result, these systems currently offer large GHG mitigation potentials (e.g., soil organic carbon (SOC) sequestration), which can be realized through the implementation of good management and sustainable agricultural practices. In this chapter we synthesize current available methodologies designed to assess SOC stocks and stock changes. From this analysis, it becomes apparent that the design and subsequent implementation of any quantification and monitoring scheme envisaged for studies focusing solely on the soil component greatly differs from those developed for whole ecosystem accounting, not just in its approach, but also in the amount of resources needed to implement it within a given degree of accuracy. We provide analyses and recommendations on methods specifically dealing with quantification and assessment of SOC at both the individual farm and the landscape scale in smallholder agricultural systems.

7.1 Introduction

Agricultural activities are responsible for about one-third of the world's greenhouse gas (GHG) emissions and this share is projected to grow, especially in developing countries (IPCC 2007). Indeed, smallholder agricultural systems are highly dynamic and heterogeneous environments that may have significantly contributed to GHG emissions over the past number of decades (Berry 2011). Furthermore, these systems traditionally suffer from severe soil organic matter (SOM) depletion due to intense decomposition following soil ploughing, the

G. Saiz (✉)
Karlsruhe Institute of Technology, Institute of Meteorology and Climate Research, Atmospheric Environmental Research (IMK-IFU), Kreuzeckbahnstrasse 19, 82467 Garmisch-Partenkirchen, Germany
e-mail: gustavo.saiz@kit.edu

A. Albrecht
Institute of Research for Development (IRD), Montpellier, France

T.S. Rosenstock et al. (eds.), *Methods for Measuring Greenhouse Gas Balances and Evaluating Mitigation Options in Smallholder Agriculture*,
DOI 10.1007/978-3-319-29794-1_7

removal of most of the aboveground biomass during harvest, and the enhanced soil erosion inherent to those activities. Yet, they may also offer large mitigation potentials through the implementation of good management and sustainable agricultural practices, particularly through improvements in land-use management, as nearly 90 % of IPCC-identified technical potential lies in enhancing soil carbon sinks (Lipper et al. 2011).

A number of methodologies are currently available for the quantification of carbon stocks in terrestrial ecosystems, varying widely in terms of accuracy, scale, and resources needed for their implementation (e.g., Pearson et al. 2005; Ravindranath and Ostwald 2008; Hairiah et al. 2010). Table 7.1 offers a comparative analysis of methods for quantification of soil organic carbon (SOC) stocks and changes with regard to level of accuracy, scale, resources demanded, and land covers considered. While nearly all the schemes feature soil as a component of the total carbon pool, the number of methods specifically designed to assess SOC stocks and stock changes are considerably more limited. This is despite the wide acknowledgement that many ecosystem services are strongly correlated with SOC levels, and their huge importance for sustaining local livelihoods. The design and implementation of any quantification and monitoring methodology for studies focusing solely on the soil component may greatly differ from those developed for whole ecosystem accounting, not only in approach or the accuracy but also in necessary resources. Therefore, it is justified to develop methods that can effectively deal with soil carbon quantification and monitoring for a given accuracy within the available budget. In the present work we focus on the soil component and provide analyses and recommendations for methods to quantify SOC in smallholder agriculture in tropical environments.

The SOC inventory in a given soil profile is controlled by the complex interaction of many factors, including climate, soil texture, topography, fire frequency, land use, and land management (Bird et al. 2001; Saiz et al. 2012). These drivers exert contrasting influences on SOC stocks at different spatial scales. At the local scale, biotic factors and management activities play a fundamental role in affecting the quantity and quality of carbon inputs and decomposition processes, while at larger scales the variation in SOC stocks is mainly controlled by topographic, edaphic, and climate-related factors (Wynn and Bird 2007; Allen et al. 2010; Saiz et al. 2012). Ultimately, an increase in SOC levels at a given site may occur either through the reduction of factors promoting SOM mineralization and lateral exports (e.g., erosion), and/or by increasing SOM inputs and enhancing stabilization mechanisms (e.g., physical protection of SOM through stable aggregates).

Given the inherent high spatial variability of SOC, accurate quantification and monitoring of SOC stocks and stock changes is a complex task even in relatively homogeneous ecosystems. This complexity is further exacerbated in smallholder environments by the existence of multiple land use activities occurring at various management intensities. Moreover, sources of uncertainty and suitable levels of precision and accuracy differ when working at the landscape scale as opposed to the farm scope because biogeochemical processes affecting SOC dynamics operate and interact at different spatial scales (Veldkamp et al. 2001; Milne et al. 2013).

Table 7.1 Comparative analysis of methods for smallholder quantification of SOC stocks and changes with regard to level of accuracy, scale, resources demanded, and land covers considered

	Accuracy		Scale		Resources demanded				Land covers considered				
Method/quantification approach	Uncertainty given	Widely tested/ used	Landscape	Farm/ paddock	Cost	Specialist expertise	Specialized equipment	Landscape stratification for sampling	Cropland	Rangeland/ grazing land	Grassland	Agroforestry	Woodland
(a) Methods specifically designed for SOC quantification													
A protocol for modeling, measurement, and monitoring soil carbon stocks in agricultural landscapes. Aynekulu et al. (2011)	x	x	x	a	MH	M	LM		x		x	x	x
VCS Module VMD0021 Estimation of Stocks in the soil carbon pool	x		x	x	H	MH	MH	x	x	x	x	x	x
Sampling, measurement and analytical protocols for carbon estimation in soil, litter, and coarse woody debris. McKenzie et al. (2000)	x	x	x	x	M	M	M	x	x	x	x	x	x
Soil testing protocols at the paddock scale for contracts and audits—Market-based instrument for soil carbon. Murphy et al. (2013)	x		x	b	LM	M	M	x		x	x		

Table 7.1 (continued)

Method/quantification approach	Accuracy		Scale		Resources demanded				Land covers considered				
	Uncertainty given	Widely tested/ used	Landscape	Farm/ paddock	Cost	Specialist expertise	Specialized equipment	Landscape stratification for sampling	Cropland	Rangeland/ grazing land	Grassland	Agroforestry	Woodland
(b) Methods considering SOC as part of whole ecosystem carbon quantification													
Measuring Carbon Stocks Across Land Use Systems (ICRAF). Hairiah et al. (2010)	x	x	x	x	LMH	LMH	LMH	x	x		x	x	x
Guide to Monitoring Carbon Storage in Forestry and Agroforestry Projects. MacDicken (1997)	x	x	x		LM	LM	LM	x				x	x
GOFC-GOLD (2009)	x		x		MH	MH	MH						x
Forest Carbon Stock Measurement: Guidelines for Measuring Carbon Stocks in Community-Managed Forests Subedi et al. (2010)	x		x		M	M	M	x				x	x
Small-Holder Agriculture Monitoring and Baseline Assessment (SHAMBA) methodology. Berry et al. (2012)	x		c	x	LM	M	M	d	x			x	
Integrating Carbon Benefits into GEF Projects. Pearson et al. (2005)	x	x	x	x	M	MH	MH	?	x	x	x	x	x
Carbon Inventory Methods. Ravindranath and Ostwald (2008)	x		x		LMH	MH	MH	?	x		x	x	x

Notes: LMH (low, medium, high)

[a]If remote sensing available

[b]Specifically designed for paddock (generally an extensive rangeland)

[c]Not designed for sites with signs of significant erosion

[d]Only areas with climate smart agricultural activities considered

Therefore, efficient sampling designs are needed across smallholder agricultural systems to ensure that SOC stocks and stock changes can be detected at various scales for a given accuracy and at minimum costs (Milne et al. 2012; Singh et al. 2013). Chapters 2 and 3 in this book provide some critical discussions on sampling designs specific to smallholder contexts. These chapters deal with systems characterization and targeting, and determination GHG emissions and removals associated with land use and land cover change.

In the present work, we propose an integrated field-based approach for small household systems that encompasses estimates of SOC stocks and stock changes both at farm and landscape scales over a wide range of land use management intensities.

7.2 Quantification of Soil Carbon Stocks

7.2.1 Sampling Design: Stratification of the Project Area

While the establishment of a geographical extent for quantification of SOC stocks and stock changes at the farm level can be straightforward, it is not the case for smallholder landscape assessment. The landscape concept may be defined by a geographic or ecological boundary, which often includes a mosaic of land covers and land uses that are managed in several different ways by the multiple stakeholders involved. In this context, Chap. 2 in this book provides recommendations for stratifying the landscape according to its agricultural productivity, economic outputs, potential GHG emissions, and social and cultural values. A SOC quantification scheme could integrate with such a stratification approach at the landscape level.

Herein, we describe the methods specifically dealing with quantification and assessment of SOC at both the individual farm and the landscape scale in smallholder agricultural systems.

Farm Level

Intensive work conducted over the past decade in smallholder agricultural systems in sub-Saharan Africa has demonstrated the existence of within-farm variability of soil fertility and related soil properties (Prudencio 1993; Carsky et al. 1998; Tittonell et al. 2005a, b, 2013). A common feature of these farming systems is the existence of strong gradients of decreasing soil fertility with increasing distance from the homestead, which mainly occur as a result of differential resource allocation driven by the farmer. This spatial gradient must be taken into account when designing SOC sampling strategies in these agricultural systems, and more so considering that previous work has also identified strong correlations between yields, soil quality indicators, land use management, and the distance from the homestead (Tittonell et al. 2005b, 2013). On the other hand, the presence of either annual or perennial vegetation on a given land use may have a strong impact on SOC stocks, as they significantly determine both the quantity and quality of organic matter inputs into the soil (Guo and Gifford 2002; Saiz et al. 2012). Therefore, distance from the homestead and land use classified by the presence of annual or perennial vegetation, are the main criteria to use in order to categorize field types for the purpose of soil sampling. Accordingly, fields are classified into home gardens, close-distance, mid-distance, and remote fields following a similar procedure as in Tittonell et al. (2005b). These areas may contain several land uses, and as it may not be feasible to sample all of them, priority should be given to the actual representativeness of the land uses being considered. Therefore, sampling should be preferentially done in the largest fields provided that management activities with potentially heavy impact on SOC stocks, such as manure additions or recurrent burning of stubble, are roughly comparable between the different land uses. However, this assumption may not hold quite true in these farming systems, and thus it is worth noting that if land use management needs to be adequately quantified, then the sampling effort may need to be increased quite considerably. Nonetheless we hypothesize that, on the whole, soil sampling across a spatial gradient may partially account for the effect of land management intensities along the farm, given that such activities are also likely to occur along the same gradient.

Landscape Level

Assessment of SOC stocks at the landscape scale can be done following a spatially stratified randomized sampling design, as this will allow for a more optimum areal coverage and unbiased assessment of sample mean, variance, and estimation variance of the sample mean. At the landscape level, the stratification can be done either through: (a) ancillary data, or (b) geographic coordinates, which may include the use of a systematic grid over the project area (de Gruijter et al. 2006).

Stratification through *ancillary variables* requires the establishment of discrete strata on which selected factors affecting SOC stocks show some degree of uniformity. Once the study boundaries have been defined, the use of remote sensing in combination with geophysical and management information may provide an effective means to stratify the target area (Ladoni et al. 2010). Such stratification needs to be performed considering, at minimum: available soil classifications, soil texture, landform information, topographic position, land cover, land use, management history, fire records, and obvious soil erosion/deposition processes. The initial stratification should be conducted in a hierarchical order whereby the factor that exerts the strongest influence on SOC stocks is ranked first, and other factors with less influence on SOC are subsequently assigned (e.g., a classical ranking approach might be climate, soil texture, land cover and management, etc.). The VCS module (VMD0018) provides detailed methodology on how to implement and adapt the stratification to the needs of the sampling process. Ideally, the number of samples to be measured in each stratum should be determined as a proportion of the area and the variance observed for that particular stratum. For this, a pilot soil sampling can be conducted which would serve a double purpose: to obtain an initial estimate of the variance for each stratum and serve as a training exercise for technicians who will be involved in subsequent sampling (MacDicken 1997). Nonetheless, it is likely that in smallholder systems, a stratum defined by biophysical factors may still be made up of land parcels managed in highly contrasting ways. Indeed, land management could account for more variation in SOC stocks at the landscape/regional level than either soil types or land use. Under such circumstances, there may be a need to stratify into a greater number of land use categories to account for land use management practices between farm tenancies (Bell and Worrall 2009). Consequently, the number of samples needed to account for spatial patterns and uncertainty in a highly heterogeneous environment can quickly become impractical due to the cost and time associated with sample collection, preparation, and analyses. To avoid this, *spatially stratified systematic sampling* approaches such as the one employed by the Land Degradation Surveillance Framework (LDSF; Aynekulu et al. 2011; Vågen et al. 2015) are easier to establish and monitor, and therefore may be a cost-effective alternative to provide a representative landscape estimate of SOC stocks and their changes. Moreover, the resulting sampling locations are spatially dispersed across the study area, but the range of variation in SOC stocks is not as effectively covered as with the stratification by ancillary variables. Therefore, the user should make his/her own choice depending on the available resources and the degree of accuracy required. We advocate the stratification by ancillary variables. However, in the case of very large heterogeneous regions, we recommend the implementation of a spatially stratified systematic sampling. It is worth stressing that while both stratification approaches (spatial and using ancillary variables) can yield relatively accurate information about SOC stocks at the landscape level, they lack proper accounting at the farm scale unless specific sampling strategies within a given household are further implemented.

The number of plots required to estimate SOC stocks in each stratum depends on the desired precision, often set at ±10 % of the mean at 90 or 95 % confidence level. The number of plots per stratum can be ascertained through the relationship described by Snedecor and Cochran (1967); See specifics in the detailed methodology section (Appendix A).

An initial soil sampling campaign should be conducted to establish baselines that can be used as references to monitor changes in SOC stocks. The level of precision required for a SOC inventory will undoubtedly influence the number of plots to be sampled, which will have necessarily a very strong impact on the cost associated with fieldwork and soil processing. Indeed, the largest component of the total cost incurred in SOC surveys corresponds to soil sampling and preparation (Aynekulu et al. 2011). Except for the case of surveys in which extremely large numbers of samples are collected (>2000), the actual cost of soil analyses is relatively low compared to the total expenditure derived from the collection and preparation of samples. Withal, and in order to minimize the number of samples to be analyzed, an extensively applied method is the bulking (pooling) of samples collected within a plot at the same depth interval. This procedure has been shown to be a cost-effective technique for smoothing out local heterogeneity and for achieving robust local and regional estimates of SOC inventories (Bird et al. 2004; Wynn et al. 2006; Saiz et al. 2012).

The specific objectives of the study shall ultimately dictate the sampling priorities, which combined with the available resources, will determine the methodology and sampling intensity to apply.

7.2.2 *Sample* Collection

(*See also the Simplified Protocol for this purpose in* Appendix B)

Ideally, samples undergoing analyses should be as representative as possible of the area of interest. To help with this, samples can be combined to provide a single representative composite sample, but there should be at least several composite samples per selected plot to provide an estimate of variance. Therefore, we propose to take three soil samples (which will be subsequently pooled by depth interval before analyses) at four locations in each plot. A plot will correspond to a given field and land use within each selected farm. The initial sampling location will roughly be allocated at the center of the field, with three replicates laid out according to a pattern of three axes separated 120° with respect to an initial axis pointing north. The replicates will be selected along these axes at approximately mid-distance between the center of the field and its boundaries. The final sampling locations will be georeferenced using a GPS, and notes should be taken about the sampling location with regard to the proximity of perennial vegetation (i.e., shrubs, trees, etc.), and any other relevant information such as presence of rock outcrops. Unless very intensive sampling is required in a given particular field, then the low analytical load proposed at the field scale (four composite samples) does not allow for proper intercomparison of small-scale intercropping, or for comparison between furrows and ridges. Therefore, sampling should be systematically allocated at the same ploughing feature (e.g., furrow).

Previous to any sampling surface litter will be removed by hand. Soil samples will then be collected at **0–10** and **10–30** cm depth intervals making use of a steel corer. This procedure will allow for determinations through the retrieval of a single soil core of both OC abundance and accurate soil bulk density (SBD) at each depth interval. Accurate determination of SBD in the topsoil layers is particularly critical given that it is at these shallow locations where SBD shows the largest variability and significantly large quantities of OC are stored. Nevertheless, it is important to note that while the use of a steel corer may be a feasible procedure in many arable lands as a result of both soil being regularly disturbed and stones being progressively removed over the years, the use of a soil auger may be necessary to collect samples in stony or very hard soils. Indeed, impenetrable layers permitting, soil sampling at **30–50** cm needs to be carried out individually at each of the four sampling locations. In this case, replication at each sampling location is avoided because of the considerable extra time and effort that would be required. Section 7.2.4 explains the different procedures that can be used to calculate SOC stocks.

7.2.3 Sample Preparation and Analytical Methods

(*See also the Simplified Protocol for this purpose in* Appendix B)

Once in the laboratory, samples are weighed in their sealed bags, clumps broken by hand and then oven dried at 40 °C to constant weight. Thereafter, an aliquot of each sample will be oven dried at 105 °C for 4 h which will allow for the calculation of SBD, while the remainder of the samples will then be dry sieved to 2 mm and gravel and root content >2 mm determined by weight.

Standard methods of soil carbon analysis such as dry combustion or wet oxidation are extensively used in SOC studies as they provide optimum quality results. Moreover, elemental (dry) combustion appliances can be coupled to mass spectrometers to provide stable isotopic carbon signatures of SOM, which broadens the possibilities for better assessing soil carbon dynamics (Bird et al. 2004). However, the elemental combustion technique is resource-demanding and may be impractical or too expensive for large sets of samples and for continuous monitoring (Aynekulu et al. 2011; Batjes 2011). Nonetheless, the amount of time required to estimate SOC stocks and the sampling and analytical costs can be greatly reduced by employing emerging techniques for in situ estimation of SOC. Among such techniques the one that has been most widely used, and thus tested, is the Infrared Reflectance Spectroscopy, either at the Near or Mid-infrared reflectance spectroscopy (NIRS or MIRS), which once calibrated can provide rapid accurate SOC estimates (Shepherd and Walsh 2002, 2007; Aynekulu et al. 2011). Despite its usefulness and versatility, it is still necessary that a significant proportion of samples (i.e., 20 %) covering the projected range of SOC values for a given inventory are analyzed using standard SOC analytical procedures. This will in turn offer the necessary calibration set to confidently apply either MIRS or NIRS to the total set of samples. The use of remote spectroscopy on airborne or satellite-mounted sensors can also provide spatially distributed and resource-efficient measurement of SOC content (Ladoni et al. 2010). However, these techniques still require simultaneous ground observations to allow for proper calibration, and there are several major challenges associated with data accuracy (Croft et al. 2012; Stevens et al. 2006).

7.2.4 Quantification of SOC Stocks

There are different approaches to account for soil carbon stocks and stock changes, and they all aim at providing a measure of mass of SOC per unit ground area.

The *spatial coordinate* approach calculates stocks considering the amount of carbon contained within a given volume of soil, which is defined by the sampled area and the depth referenced to the surface level. With this approach, the average SOC stock for a given depth interval (d) is calculated according to the following formula:

$$\mu_{\mathrm{d}} = \mathrm{BD}_{\mathrm{d}} \times \mathrm{OC}_{\mathrm{d}} \times \mathrm{D} \times (1 - \mathrm{gr}) / 10;$$

where:

μ_d is SOC stock (Mg OC ha^{-1})
BD_d is soil bulk density (g cm^{-3})
OC_d is the concentration of OC in soil (<2 mm; mg OC g^{-1}soil)
D is soil depth interval (cm)
gr is fractional gravel content, the soil fraction >2 mm

However, the amount of soil contained within a given volume (SBD) may change as a result of swelling and/or compaction caused by land use change or management. Under those circumstances, sampling to a fixed depth from the surface (spatial coordinate approach) will result in different amounts of soil mass being sampled for the same volume, while the soil C concentration per unit dry soil mass might not have changed. This can lead to errors in the interpretation of changes in SOC storage following disturbance.

The determination of SOC stocks can also be achieved through *cumulative* or *material mass coordinate* approach, which consists of collection and quantification of all the soil mass in a given depth interval. The use of *cumulative mass coordinate* approach is widely used to correct for differences in bulk density that may have been caused by land use change or agricultural practices. Moreover, the adoption of this method may improve our ability to make comparative measurements across time, treatments, locations, and equipment (McKenzie et al. 2000; Gifford and Roderick 2003; Wuest 2009). Furthermore, since sampling by mass avoids potential biases derived from varying bulk density caused by land use change or agricultural practices, it is often regarded as the method of choice for SOC monitoring over time (see McKenzie et al. 2000 and Gifford and Roderick 2003 for detailed guidance on the method). Nonetheless, compared to soil coring, this method requires additional effort and skill. In the cumulative mass approach, depth varies such that each sample contains the same dry mass per unit ground area. Gifford and Roderick (2003) provide in-detail explanations and examples on how to determine SOC stocks using this methodology. Briefly, the method involves coring a bit deeper than the nominal depth involved (e.g., 55 cm for a required 50 cm depth) and each full soil core is then divided into several segments. We recommend sampling at 10, 30, 50, and 55 cm in those cases where coring may be feasible in order to compute for SBD and be able to interconvert between the spatial coordinate and the cumulative mass coordinate approach.

Another method that has been recommended to quantify SOC stock changes is the *equivalent soil mass* approach (Ellert and Bettany 1995; Lee et al. 2009). It consists of correcting for differences in SBD through the calculation of the mass of SOC in an equivalent soil mass per unit area (i.e., the heaviest soil layer is designated as the equivalent mass, against which to calculate the thickness of the soil that is required to obtain such mass). However, its implementation is even more difficult than the coordinate mass approach (McBratney and Minasny 2010).

Regardless of the method used to quantify SOC stocks, the provision of SBD data is of great importance so as to understand and interpret SOC dynamics (Gifford and Roderick 2003). In the case of soil augering, the calculation of SBD can be achieved by sand-filling the auger-hole volume. Alternatively, one can use soil density rings, which are orthogonally inserted onto the wall of a dugout soil pit. These are however highly time consuming as well as demanding tasks, and hence they should be limited to cases in which coring is not possible.

7.2.5 Scaling SOC Stocks to Landscape and Whole Farms

There is a lack of standardized methodologies to scale up SOC stocks from a point source (pedon) to regional (landscape) and larger spatial scales. In this work, the scaling up of SOC stocks at the landscape scale is achieved through the proposed spatially stratified randomized sampling design. Accordingly, the average SOC stock for a given stratum is calculated as follows:

$$\mu_{\mathrm{st}} = \frac{1}{n}\sum_{i=1}^{n} y_i;$$

where:

μ_{st} is the mean SOC stock for stratum st
y_i represents each calculated SOC stock in that stratum
n is the number of observations in that stratum (see Appendix A for detailed calculations on the number of plots required in each stratum)

The *variance* in SOC stocks for a given stratum is calculated according to the following formula:

$$\sigma_{\mathrm{st}}^2 = \frac{1}{n-1}\sum_{i=1}^{n}\left(y_i - \mu_{\mathrm{st}}\right)^2;$$

where:

σ is the SOC stocks variance
y_i represents each calculated SOC stock in that stratum
μ_{st} is mean SOC stock associated with the stratum st
n is the number of observations in that stratum

The *average* SOC stock for the area of study (landscape) is calculated considering both the mean SOC stock obtained for each stratum and the area occupied by each stratum. Therefore, the calculation is as follows:

$$\mu = \frac{\sum_h^H a_h \times \mu_h}{A};$$

where:

μ is the mean SOC stock
a_h is the area of the stratum h
μ_h is mean SOC stock associated with the stratum h
A is the total area of the study

The average *standard error* in SOC stocks for the area of study (landscape) is calculated according to the following formula:

$$\mathrm{SE} = \sqrt{\sum_{h=1}^{H} \left(\frac{a_h}{A}\right)^2 \times \frac{S_h^{\ 2}}{a_h}};$$

where:

SE *is the standard error for the entire population*
a_h is the area of the stratum h
S_h is the variance of stratum h
A is the total area of the study

Scaling SOC stocks from a few point source measurements (fields) to the whole farm necessarily requires a series of assumptions unless all fields within the farm are sampled (which may be highly unpractical). Here, it is assumed that the center and perimeter of each field are georeferenced so that the field's surface area can be determined. In the proposed scheme, samples within a given farm should be taken along the previously described land use intensity gradient (i.e., home gardens, close-distance, mid-distance, and remote fields) at their most spatially representative fields. If for a given section (i.e., close-distance fields), there is an occurrence of individual fields with annual and perennial vegetation (crops or trees), and the area of the smaller field is at least half the size of bigger field, then sampling should be conducted at both fields. The *average* SOC stock for the selected farm is then calculated considering both the mean SOC stock obtained for each section and the area occupied by each section. The calculation procedure is similar to the one described for the landscape scale, and it simply replaces strata by sections.

Uncertainties in SOC stock assessments vary according to the scale and the spatial landscape unit. Goidts et al. (2009) demonstrated that scaling up field scale measurements to the landscape level increases the coefficient of variation of SOC estimates. However, the same work showed that such uncertainty may be smaller than errors associated to local spatial heterogeneity and analytical procedures.

7.3 Quantification of Soil Carbon Stock Changes

The determination of the sampling intensity required to demonstrate a minimum detectable difference in SOC stocks over time has been the subject of numerous studies (Garten and Wullschleger 1999; Conen et al. 2004; Smith 2004). The actual number of samples to detect SOC differences for different degrees of statistical confidence will be directly dependent on the background level that the study requires (i.e., the detectable difference in SOC relative to the stock baseline estimated in the first inventory). Moreover, considering the inherent natural variability of soil properties, the demonstration of small changes in SOC stocks may often require the collection of an impractically large number of samples (Garten and Wullschleger 1999), whose costs may quickly overrun any financial benefit derived from a potential increase in SOC levels. Therefore, different approaches have been used to monitor SOC stock changes, which invariably represent a compromise between accuracy and cost. Table 7.2 shows a comparison of methods used to monitor SOC stock changes classified according to the level of accuracy, scale, and resources demanded.

7.3.1 Repeated measurements

A further classification is made on the basis of the measurement domain (where the analyses take place).

Laboratory-Based Analyses

These are the most widely used techniques, which involve physical collection and subsequent processing of soil samples (see Sect. 7.2.3). The standard methods used for soil carbon analysis are dry combustion, wet oxidation, and the use of reflectance spectroscopy, which is increasingly being used over the past number of years as an effective way to optimize time and analytical costs. However, some controversy still exists about the compatibility of data derived from different spectroradiometers (Reeves 2010), and there is still a need for collection and analyses by conventional techniques of a significant proportion of samples to allow for calibration of the entire sample set.

In Situ Analyses

While lab-based analyses provide high-quality results, they are resource-demanding and may be impractical or too expensive for continuous monitoring of SOC (Aynekulu et al. 2011; Batjes 2011). The implementation of SOC analyses in the field by means of portable spectroscopy allows for the assessment of a much larger number of sampling locations compared to that offered by lab-based methods, as the former is a fast, cost-effective, and non-destructive technique. However, its accuracy is lower than that provided by conventional methods, since there are issues related to soil surface conditions such as soil moisture and surface roughness, which

Table 7.2 Comparison of methods for monitoring SOC stock changes with regard to level of accuracy, scale, and resources demanded

			Accuracy			Scale			Resources demanded		
Quantification approach	Measurement domain	Measuring technique	Analytical precision	Need for calibration	Widely tested/ used	Spatial coverage	Landscape	Farm	Cost/ time	Specialist expertise	Specialized equipment
Repeated Measurements	Laboratory-based analyses	Conventional analyses (e.g., dry combustion, wet chemistry)	H		H	L	L	M	H	M	M
		Lab-spectroscopy	M-H	X	M-H	L-M	L	M	M-H	M-H	M
	In situ analyses	Ground-based spectroscopy	M-H	X	L-M	M	M	M-H	M	M-H	M
	Remote spectroscopy	Processing of spectroscopic imagery	M	X	L-M	H	H	c	L-M	H	M-H
Modeling	Simulations	Process-based models of SOM dynamics (e.g., Century, Roth-C, DNDC)	L-M	X	a	d			L	H	L
		IPCC Good Practice Guidance (LULUCF)	L		b				L	L	L

Notes: LMH (low, medium, high)

[a] See text for limitations and assumptions needed when using modeling

[b] Emission factors provided by IPCC Good Practice Guidelines to calculate changes in SOC stocks have been widely used, but due to its inherent generality is an extremely coarse instrument for proper quantification of changes in small household systems, and it is mentioned here for reference purposes only

[c] The use of remote spectroscopy at the farm scale is strongly limited by the resolution of the available imagery

[d] Proper simulation of SOC dynamics at the landscape and farm scale may be possible provided it is conducted on sites with accurate management information

may affect the spectral signal. Therefore, there is a need to conduct a statistical calibration before each field campaign in order to achieve an acceptable level of accuracy (Stevens et al. 2006).

Remote Spectroscopy

The use of reflectance spectroscopy on airborne or satellite-mounted sensors provide high temporal resolution and allow for an improved representation of the spatial variation of SOC in a cost-efficient manner (Ladoni et al. 2010; Croft et al. 2012; Stevens et al. 2006). Nonetheless, there are still major constraints with regard to using this technique as a plausible method to detect SOC stock changes. Croft et al. (2012) highlight some of these limitations, which include: the comparatively higher analytical uncertainty than that obtained from conventional or ground-based reflectance spectroscopy; the high spatiotemporal variability of soil surface conditions that can affect the spectral signal (e.g., soil moisture, vegetation or crop residue cover, differences in soil surface roughness, etc.); the spatial uncertainties associated with instrument spatial resolution and SOC spatial heterogeneity; and the need for atmospheric correction and simultaneous ground data collection to calibrate and validate the output of such studies. Furthermore, remote spectroscopy can only use the reflectance of bare surface to measure soil properties and is not able to detect vertical gradients in SOC within the topsoil (Stevens et al. 2006). Finally, there is a dearth of studies using remote spectroscopy in arid or semi-arid regions, which host a large amount of small household farming systems. In these environments SOC contents are typically low and the interference with other soil properties (e.g., $CaCO_3$ or $CaSO_4$ contents) may change the spectral behavior of soil considerably, which could have further detrimental effects on the performance of the remote sensing techniques (Ladoni et al. 2010). Withal, the detection limit of these techniques is still too high to use them for SOC stock change studies (Stevens et al. 2006). To make these techniques fully operative, additional efforts must be taken to decrease the detection limit.

7.3.2 Modeling

Compared to measuring techniques that require the implementation of repeated measurements to quantify SOC stock changes, the use of process-based models (e.g., DNDC, Roth-C, Century) have obvious advantages in terms of resources demanded. Moreover, models can provide relatively fast and inexpensive simulations of SOM dynamics at different spatiotemporal scales. However, such simulations are based on a number of assumptions that will necessarily result in very large uncertainties of the estimates obtained. Here, we briefly describe some of the main weaknesses of models that could potentially be used to quantify SOC stock changes within the context of small household agricultural systems in tropical environments.

Assumption of Stable Conditions

Most SOM dynamic models assume stable conditions in SOM pools prior to modeling how factors like management or climate change affect their dynamics. However, the vast majority of small household systems in the tropics are not necessarily in steady state conditions. In the tropics, large tracts of land under current agricultural practices have been covered by natural ecosystems not much longer than a generation ago, but in many cases this would only be a few decades or even just some years ago (Houghton 1994; FAO and JRC 2012). Because of this, current SOM dynamics will still be highly influenced by past vegetation. Therefore, the assumption of stable conditions in those systems is likely to result in gross inaccuracies. While the influence of past vegetation might of course be modeled, this would be done at the expense of bringing on further uncertainty to the results, as this impact is likely to vary with the type of vegetation, time since conversion, landscape position, soil type, etc.

Coupling Erosion Processes

Quite a significant number of small household systems are established on slopes of varying degrees, with farms being increasingly established on steep marginal land as a result of population pressure. Moreover, cropped fields may be void of vegetation for some time during the year, or in some cases, the entire year (fallow). The combination of those factors makes soil erosion a highly significant factor, which may naturally lead to lateral transfers of SOM. Again, coupling a soil erosion model to a SOM dynamic one can be attempted, but the resultant application would need to be parameterised for the wide array of heterogeneous conditions existing between farm managements, the different land uses, soil types, etc., all of which may undoubtedly produce an even greater source of uncertainty.

Existence of Contrasting SOM Dynamics Between Crops

Small household systems are highly dynamic in terms of the crops being used (C_3 plants such as legumes and napier grass; and C_4 plants such maize and sorghum) whose presence and abundance may vary between years within the fields of a given farm. There is increasing evidence that C_3 and C_4 vegetation have a strong influence on SOM processes, see for instance Wynn and Bird (2007) and more recently Saiz et al. (2015). Besides inherent microbial processes and material (biomass) recalcitrance, these dynamics are highly influenced by soil texture through their effect on abiotic properties. Therefore, vegetation may exert very strong effects on SOC stocks, which traditional SOM dynamic models are not yet able to simulate.

In summary, models can provide very useful indications about trends of SOM levels with respect to changes in climate and/or management, and they can do so at high spatiotemporal resolutions and at a fraction of the cost of those using repeated measurements (Table 7.2). However, the uncertainties associated to the estimates are currently too large to use them as a verifiable tool to demonstrate SOC stock changes, particularly in these highly heterogeneous systems. At the very least, models require high-quality data gathered at different time intervals for proper parameterisation, and this is still an important aspect clearly lacking for these grossly understudied tropical systems (Rosenstock et al. 2016, Chap. 9).

7.3.3 Monitoring Frequency and Recommendations

While IPCC (2003) and IPCC (2006) recommend 5- and 10–20 year monitoring intervals respectively, a relevant sampling interval suited to site-specific conditions can be ascertained by using models of SOC dynamics to plan both the frequency and intensity of subsequent surveys for determining SOC stock changes (Smith 2004). However, modeling of highly heterogeneous environments such small household agricultural systems in tropical systems is a challenging task, which is unlikely to provide a single answer with regard to when and how intensively different sites should be measured to detect significant changes in SOC stocks. Alternatively, estimation of changes in SOC over shorter periods could be achieved through the measurement of changes in particular soil carbon fractions (e.g., particulate organic matter) given that these are more sensitive to changes than total carbon in the bulk soil (Six et al. 2002). While this is a rather useful qualitative assessment of SOC sequestration it does not reflect the overall SOC stock changes that should be simultaneously assessed, thus increasing the overall cost and sampling effort. Furthermore, the implementation of a SOM fractionation procedure requires specific laboratory equipment (i.e., sonicator) and access to relatively expensive consumables (i.e., heavy liquid; Wurster et al. 2010).

We recommend adopting a strategy similar to the one proposed by Lark (2009), which suggests sampling only a proportion of the initial baseline sites in any one stratum. This strategy purposely focuses efforts in those locations likely to show the larger differences in SOC stocks over a fixed term (i.e., 10-year period). Thereafter, the strata that show a large change could then be sampled more intensively. Locations likely to show the larger changes in SOC stocks will normally include fields affected by intensive management, those having changed land use since the last survey, and the ones presenting recent signs of land degradation. We also advise pairing sampling locations in space as this may allow for a more effective detection of SOC changes in time (Ellert et al. 2007), and a sampling scheme consistent with that used in the first round of sampling. Furthermore, collection of samples should be routinely conducted at roughly the same time of the year, and in between relevant agricultural practices (i.e., harvesting, fertilization, etc.). Further information about quantifying SOC over time is given in the Appendix A.

We would like to conclude this section on SOC stock changes stressing that the only way to detect reliable signals and early trends in soil monitoring schemes is to improve the overall measurement quality (precision and bias) and to shorten the measurement periodicity (Desaules et al. 2010). However, the labor, analytical costs, and time needed to achieve a given sensitivity might overrun the potential monetary benefits derived from a hypothetical increase in SOC levels. As an illustrative case, Smith et al. (2001) indicate that between 10 and 20 samples should be collected to detect a 15 % change in SOC stocks in a relatively homogeneous system (<25 % coefficient of variation). Moreover, special attention should also be placed on the issue of permanence as most of the new SOC fixed as a result of improved management activities is in a labile form (particulate organic carbon), and thus, it is highly prone to be lost back to the atmosphere in a relatively short timeframe if conditions changed. Therefore, emphasis should be placed on promoting sustainable agricultural practices, as these will bring both economic and environmental benefits to the farmers in the medium term. Enhanced SOC sequestration may indeed be one of those benefits, but in our view it should not be the purpose of grand resource-demanding monitoring schemes, especially if the time elapsed between surveys has not been long enough (i.e., at least 10–20 years). Bearing this in mind, and even considering that at present proper simulation of SOM dynamics is very limited in small household systems because of the scarcity of high-quality data, modeling still represents an alternative that, provided high-quality data was available, could be applied across broad spatiotemporal scales in a cost-effective manner. Therefore, we propose the establishment of permanent monitoring sites across a gradient of management qualities (from highly intense to poor management scenarios) in the geographical area of interest to serve as reference sites to generate data that can be used for model parameterization and validation for farming practices under small household conditions.

Appendix A: Methodology for Quantification of Soil Carbon Stocks and Carbon Stock Changes

Number of Plots Required

The number of plots required to estimate SOC stocks in each defined stratum depends on the desired precision, often set at ±10 % of the mean at 90 or 95 % confidence level. In the case of strata defined by ancillary variables, the number of plots per stratum can be ascertained through the relationship described by Snedecor and Cochran (1967);

$$n = \left(\frac{t_\alpha S}{D} \right)^2;$$

where:

t_α is Student's t with degrees of freedom at either 0.95 or 0.90 probability level
S and D are the standard deviation and the specified error limit respectively for values obtained from an initial assessment of the stratum

On the other hand, and for the case of a given area stratified by geographical coordinates or ancillary variables, the number of plots required could be determined using a slightly modified relationship (Pearson et al. 2005; Aynekulu et al. 2011);

$$\mathrm{n} = \frac{(N \times S)^2}{\frac{N^2 \times D^2}{t_\alpha^{\ 2}} + (N \times S^2)};$$

where:

t_α, S, and D are as above and derive from values obtained from an initial assessment of the area considered
N is the number of sample units in the population, that is the total area divided by plot size

The resultant number of plots can be further allocated into a number of defined strata by using:

$$\mathrm{n}_h = \frac{N_h \times S_h}{\sum_{h=n}^{L} N_h \times S_h} \times n;$$

where:

N_h is the area of the stratum h
S_h is the standard deviation of stratum h
L is the number of strata
n is the total number of plots

In the cases where the confidence interval exceeds ±10 % with 90 % confidence, the user may undertake one of three actions (VCS module VMD0018): (a) re-stratify according to any significant correlation observed between the sample variance to geographic or other factors, (b) Increase the number of plots, and (c) set lower confidence intervals, increasing thus the estimates uncertainty. The determination of the number of plots to be sampled in each stratum as a proportion of both its area and the observed variance may certainly be an efficient approach. Adding to this efficiency, it can also be expected that the number of plots required for determination of SOC stocks for a given stratum defined by ancillary variables may be significantly small compared to the ones needed in the less homogeneous strata defined by geographical coordinates.

With regard to the number of samples required to demonstrate a given minimum detectable difference in SOC stocks over time the reader is referred to Garten and Wullschleger (1999), Conen et al. (2004) and Smith (2004) for sound descriptions of the methods and equations used. Finally, a very recent report by Chappell et al. (2013) provides excellent advice on a generic monitoring design to detect changes in SOC, which includes illustrative examples with step-by-step calculations.

Appendix B: Simplified Protocol for Taking and Processing Soil Samples, Adapted for the SAMPLES Project

This protocol covers both the soil sampling procedure and sampling processing and assumes the plots to be sampled have already been pre-selected.

Soil Sampling

Soil samples are collected in four different locations within the plot of choice to account for the inherent heterogeneity of SOC. Start roughly at the center of the plot/subplot (replicate 1) and establish the other three replicates laid out according to a pattern of three axes separated 120° with respect to an initial axis pointing north. Make sure the other three replicates are set up at a prudent distance from the edges of the plot/subplot (+5 m if possible) to avoid any boundary effects, but do try to cover ground. The final sampling locations will be georeferenced using a GPS.

It is assumed that a stainless steel corer, a soil auger, and/or a spade will be used for retrieving the samples. All samples will be placed in labeled zip-lock bags. It is very important that the bags are clearly labeled with a permanent marker. Always a good idea to label them immediately after you take the sample otherwise they may get mixed up (if a marker is not around, write it in a paper and put it inside of each bag). A good labeling should mention at the very least:

- Plot/subplot name or number (e.g., DCR)
- Replicate number (e.g., 3 for replicate 3)
- Depth (i.e., 10–30)

Then in the same bag and line in big clear letters following the example given it should say: *DCR-3* (*10–30*)

Detailed Sampling Procedure

In the case of the 0–10 and 10–30 cm intervals, three individual samples within 1 m radius will be collected. This is done to better account for local heterogeneity, which is particularly pronounced at this shallow depth. Subsequently samples from the same location and depth interval are pooled to minimize analytical costs.

0–10 cm

- Remove vegetation and surface litter.
- Push short corer (steel cylinder) into soil until the 10 cm mark is reached. Retrieve it gently by carefully shaking it back and forth sideways to compact a bit the surrounding soil (this will get subsequent sampling at depth much easier and will avoid soil crumbling into the hole).
- Pull the corer out rotating carefully (always clockwise as this will be very relevant when using the other soil sampling gear at depth).
- Place the soil into plastic bag, trying not to touch it with the hands. Starting with the topside (loose crumbly soil gets out first), and then turn the cylinder upside-down.
- To help the soil come out, use the rubber mallet to impact the cylinder walls while it gets turned around. The soil will come out eventually. Get all the soil out of the tube.

10–30 cm

- Hammer the next sampling cylinder into the soil until the depth markings. You may be using a regular cylinder or the one with a detachable cutting edge (preferably the latter as it is more robust). If using the latter, then you will have to carefully detach this cutter and scrap the sample out onto the bag. This can be done by a second person, thus improving sampling speed. Regardless, beware of what you are using as the diameters (crucial for bulk density determination) change for each choice.
- Shake it back and forth carefully sideways (to compact surrounding soil).
- Rotate clockwise, pull out and extract soil sample (using the sample extruder if using cylinders without detachable cutting edge).
- Again: put the soil into a labeled plastic bag avoiding contact with the hands.

30–50 (55) cm

- If the soil is relatively soft and free of stones, use the cylinders (as bulk density can still be used). If that is not the case, then use the soil auger or spade.
- If you are using the cumulative mass coordinate system to calculate SOC stocks, then you will need to collect all of the soil at the suggested depth intervals plus an extra one a bit deeper (50–55 cm).
- If using a spade to reach the required depth, the sample will be obtained by scratching the soil out of the walls. Prior to obtaining any sample, the walls of this hole (pit) need to be cleaned (scratched) to avoid contamination. Start

scratching/ sampling from the bottom once the hole has been finished. Take roughly the same amount of soil material along the targeted depth interval, as you do not want to take most of your sample at a concentrated point. It would be good to have a graduated ruler or stick with depth marks.

In general, also consider the following:

- Take notes that may help you to interpret results later on (GPS, land use history, farmers' comments on management, type of soil, current vegetation, evidence of erosion, fire, etc…).
- If using coring, a sample that comes broken in the first 30 cm (as a result of coarse stones/roots) cannot be used. Sampling has to be done again in another location nearby.
- Beware that if the soil is very rocky, there will be a risk of overestimating SOC stocks if using soil coring. Therefore, an estimate of rock content for a given plot should be given. However, an accurate quantification of rockiness is a very demanding task, as it would involve to purposely sampling several pits at each studied field.
- Always take note of what corer you are using (because of diameters!). It may be that you are exchanging between cylinders with different sizes for whatever logistical reason (e.g., cylinder with detachable cutting hoe vs. normal cylinder. These two have different diameters and will definitely affect bulk density calculations). This is very important, take notes.
- Be careful that the sampling hole does not get contaminated while taking samples, e.g., do not step on the hole, do not let litter or surface soil fall in, etc.
- After sampling a plot, the coring cylinders and scraper need to be thoroughly cleaned (have wet cloths with you).
- If using the auger, use the same depth intervals as those with the cores (0, 10, 30, 50).
- From the outside of the plastic bag, crumble by hand big clumps of soil into smaller parts, which will be critical for easy soil processing later on.
- Closure of bags: rolling them up releasing air from the bag and then close it, so that it contains as little air as possible.
- Take several pictures of the plot/subplot.
- As a matter of good practice, do try to fill in sampling holes with any excess soil derived from your digging.
- To calculate SBD using the auger, you will need to calculate the volume by filling the hole with sand, which is highly demanding and slow procedure.

Soil Bulk Density Determinations

In all cases, calculation of SBD should include fractions >2 mm. So before any sieving takes place the following should be done:

As soon as possible, and certainly before 2 days after collection from the field, let the samples air-dry (after opening and rolling down bags) in a rain-protected location. It is always a good idea to progressively (each day) break the soil clumps with your fingers while the bags are being dried (but be gentle or you may break the bag). A bit everyday is the best, otherwise you will find handling of samples much harder in the coming days, and will have to use a hammer. Also, avoid cross-contamination between samples by doing it from the outside of the bag (gently squeezing it with your fingers). When an oven becomes available, put the bags inside at 40 °C. After a number of days, when samples are seemingly dry (5–7 days will be safe—but of course it all depends on initial moisture content), take them out of the oven and **weigh each sample** (including the plastic bag) but wait about half an hour after the samples have been taken out to do this weighing.

After this weighing, take an aliquot of each sample and place them in labeled paper bags (about ~ ¼, of the total sample, **but weigh how much exactly before you put them inside the oven**). Dry them at 105 °C for 24 h. As before, **weigh all the samples after about half an hour after they were taken out of the 105° oven**. Once the weights of these aliquots have been recorded you can throw this material away.

In total **you should have three weights for each sample** (i.e., total soil weight, sample before oven dried at 105 °C, sample after oven dried at 105 °C). This will allow for proper calculation of SBD.

In general, also consider the following:

- Make sure you always take weights knowing which bag you are using as each different type of bag will have different weight (both plastic and paper).
- Get an average weight of five bags of each type you use, so that can be deducted from the calculations later on.
- Let the samples dry by air (open plastic bag) and roll them down.
- Always check that the oven works well.
- Let the samples cool down at room temperature for at least 30 min, unless there was a desiccator that could be used for storing samples prior to weighing. In such case, then the weighing should occur immediately after extracting the samples from the desiccator.
- Weigh the soil with its bag. Very Important!
- Balance/scale should be precise up to 0.1 g.

Sample Processing

Sieving: The remaining of each sample dried at 40° (most of it) needs to be weighed again and sieved to 2 mm. Gravel and root content >2 mm will be weighed separately. Therefore we will get the fractions of coarse roots and gravel. But first remove carefully all large clumps with a rolling pin (bakery). Removing the soil from the bag to break up any clumps is very time consuming and may lead to gross errors. Therefore, it is good practice to progressively break clumps from outside of the bag as the sample dries. After sieving, you should have three weighs for each sample (bag) in total (i.e., total soil weight, roots>2 mm, and gravel (>2 mm)).

Pooling/bulking: There are numerous ways of pooling, and the final choice depends on the purpose and load of work that can be undertaken. The methods explained below are just two ways that lead to fewer analyses to be undertaken and cover two different purposes:

1. If the aim is to get bulk soil samples to undertake just a single analyses at each plot/subplot bulk by depth interval (e.g., all samples from the same plot/subplot collected at 10–30 cm), then do as follows:

   - Use the same weight for all the replicates (20 or 30 g), and put them together in a bowl or tray. Do not use the entire sample from each bag! Keep them as back ups.
   - Mix them a bit always with clean, dry hands (10 s should be alright).
   - Put the mixture in a new bag with the same code as before but indicating "Bulk" at the end.
   - If the aim is to also get **a "master soil sample" 0–30 cm** for subsequent analyses (texture, mineralogy, organic matter fractionation, ECEC, etc.) then from the previous bulked bags the weights that need to be put together are calculated as follows:

   First the average bulk density for the Master (BDM) is calculated:

$$\text{BDM} = \left[\text{BD}_{(0-10)} \times (1/3)\right] + \left[\text{BD}_{(10-30)} \times (2/3)\right]$$

   Then to obtain about 90 g of Master sample, proceed as follows:

$$30\ \text{g} \times \text{BD}_{(0-10)} / \text{BDM}$$

$$60\ \text{g} \times \text{BD}_{(10-30)} / \text{BDM}$$

   These weights are put in a separate bag, which is to be called "master" with same code as before and indicating (0–30) at the end of the labeling.

2. Sometimes it may be necessary to have an extra bag with about 20 g of Master soil (0–30) that will be used for soil textural analyses. Take about 20 g from this bag and put them into a small bag with the same coding indicating that is for "texture."

   Powdering: If powdering is needed, then proceed as follows:

- Take about 3 g of your sieved, pooled/bulked sample.
- Powder the sample with the aid of a mortar-pestle or micromill device.
- Put the sample into a small plastic bag with the code on it.

- Be very careful that all instruments used for powdering get properly cleaned (if using water then it is very important that everything is absolutely dry again—or subsequent analyses involving weighing of the sample will be biased).
- Finally, about 50 g of sample per bag should be stored for any further potential analyses.

References

Allen DE, Pringle MJ, Page KL, Dalal RC (2010) A review of sampling designs for the measurement of soil organic carbon in Australian grazing lands. Rangeland J 32:227–246

Aynekulu E, Vågen T-G, Shephard K, Winowiecki L (2011) A protocol for modeling, measurement and monitoring soil carbon stocks in agricultural landscapes. Version 1.1. World Agroforestry Centre, Nairobi

Batjes NH (2011) Research needs for monitoring, reporting and verifying soil carbon benefits in sustainable land management and GHG mitigation projects. In: De Brogniez D, Mayaux P, Montanarella L (eds) Monitoring, reporting and verification systems for carbon in soils and vegetation in African, Caribbean and Pacific countries. European Commission, Joint Research Center, Brussels, pp 27–39

Bell MJ, Worrall F (2009) Estimating a region's soil organic carbon baseline: the undervalued role of land-management. Geoderma 152:74–84

Berry N (2011) Whole farm carbon accounting by smallholders, lessons from Plan Vivo Projects. Presentation at the Smallholder mitigation: whole farm and landscape accounting workshop. FAO, Rome, 27–28 Oct 2011. www.fao.org/climatechange/micca/72532/en/

Berry N, Cross A, Hagdorn M, Ryan C (2012) The Small-Holder Agriculture Monitoring and Baseline Assessment (SHAMBA). Tropical land use research group. School of GeoSciences. University of Edinburgh, Scotland

Bird M, Santrucková H, Lloyd J, Veenendaal EM (2001). Global soil organic carbon pool. In E.-D. Schulze, SP Harrison, M. Heimann, EA Holland, J. Lloyd, IC Prentice, et al. (Eds.), Global biogeochemical cycles in the climate system (pp. 185–197). San Diego: Academic Press

Bird MI, Veenendaal EM, Lloyd J (2004) Soil carbon inventories and d^{13}C along a moisture gradient in Botswana. Glob Chang Biol 10:342–349

Carsky RJ, Jagtap S, Tain G, Sanginga N, Vanlauwe B (1998) Maintenance of soil organic matter and N supply in the moist savanna zone of West Africa. In: Lal R (ed) Soil quality and agricultural sustainability. Ann Arbor Press, Chelsea, pp 223–236

Chappell A, Baldock J, Rossel RV (2013) Sampling soil organic carbon to detect change over time. Report to Grains Research and Development Corp and Australian Department of Environment, Australia

Conen F, Zerva A, Arrouays D et al (2004) The carbon balance of forest soils: detectability of changes in soil carbon stocks in temperate and boreal forests. In: Griffith H, Jarvis PG (eds) The carbon balance of forest biomes. Bios Scientific Press, London

Croft H, Kuhn NJ, Anderson K (2012) On the use of remote sensing techniques for monitoring spatio-temporal soil organic carbon dynamics in agricultural systems. Catena 94:64–74

de Gruijter JJ, Brus DJ, Bierkens MFP, Knotters M (2006) Sampling for natural resource monitoring. Springer, New York

Desaules A, Ammann S, Schwab P (2010) Advances in long-term soil-pollution monitoring of Switzerland. J Plant Nutr Soil Sci 173:525–535

Ellert BH, Bettany JR (1995) Calculation of organic matter and nutrients stored in soils under contrasting management regimes. Can J Soil Sci 75:529–538

Ellert BH, Janzen HH, VandenBygaart AJ, Bremer E (2007) Measuring change in soil organic carbon storage. In: Carter MR, Gregorich EG (eds) Soil sampling and methods of analysis, 2nd edn. CRC, Boca Raton, pp 25–38

FAO, JRC (2012) Global forest land-use change 1990–2005. In: Lindquist EJ, D'Annunzio R, Gerrand A, MacDicken K, Achard F, Beuchle R, Brink A, Eva HD, Mayaux P, San-Miguel-Ayanz J, Stibig H-J (eds) FAO Forestry Paper no. 169. Food and Agriculture Organization of the United Nations and European Commission Joint Research Centre. FAO, Rome

Garten CT, Wullschleger SD (1999) Soil carbon inventories under a bioenergy crop (Switchgrass): measurement limitations. J Environ Qual 28:1359–1365

Gifford RM, Roderick ML (2003) Soil carbon stocks and bulk density: spatial or cumulative mass coordinates as a basis of expression? Glob Chang Biol 9:1507–1514

GOFC-GOLD (2009) A sourcebook of methods and procedures for monitoring and reporting anthropogenic gas emissions and removals caused by deforestation, gains and losses of carbon stocks in forests remaining forests, and forestation. GOFC-GOLD Report version COP15-1. p 197 GOFC-GOLD Project Office, Natural Resources Canada, Alberta

Goidts E, Van Wesemael B, Crucifix M (2009) Magnitude and sources of uncertainties in soil organic carbon (SOC) stock assessments at various scales. Eur J Soil Sci 60:723–739

Guo LB, Gifford RM (2002) Soil carbon stocks and land use change: a meta analysis. Glob Chang Biol 8:345–360

Hairiah K, Dewi S, Agus F, Velarde S, Ekadinata A, Rahayu S, van Noordwijk M (2010) Measuring carbon stocks across land use systems: a manual. World Agroforestry Centre (ICRAF), SEA Regional Office, Bogor

Houghton RA (1994) The worldwide extent of land-use change. BioScience 44(5):305–313

IPCC (2003) Good practice guidance for land use, land-use change and forestry. Penman, J and Gytarsky, M and Hiraishi, T and Krug, T and Kurger, D and Pipatti, R and Buendia, L and Miwa, K and Ngara, T and Tanabe, K and Wagner, F (Eds). Published by Institute of Global Environmental Strategies (IGES), on behalf of the Intergovernmental Panel on Climate Change (IPCC): Hayama, Japan. ISBN: 4-88788-003-0

IPCC (2006) 2006 IPCC Guidelines for National Greenhouse Gas Inventories, Prepared by the National Greenhouse Gas Inventories Programme, volume 4, Agriculture, Forestry and Other Land Use. Eggleston S, Buendia L, Miwa K, Ngara T, Tanabe K (eds) Published by Institute of Global Environmental Strategies (IGES), on behalf of the Intergovernmental Panel on Climate Change (IPCC), Japan

IPCC (2007) Agriculture, in climate change (2007): mitigation. Working Group III Contribution to the fourth assessment report of the Intergovernmental Panel on Climate Change. Cambridge University Press, New York

Ladoni M, Bahrami HA, Alavipanah SK, Norouzi AA (2010) Estimating soil organic carbon from soil reflectance: a review. Precis Agric 11:82–99

Lark RM (2009) Estimating the regional mean status and change of soil properties: two distinct objectives for soil survey. Eur J Soil Sci 60:748–756

Lee J, Hopmans JW, Rolston DE, Baer SG, Six J (2009) Determining soil carbon stock changes: simple bulk density corrections fail. Agr Ecosyst Environ 134:251–256

Lipper L, Neves B, Wilkes A, Tennigkeit T, Gerber P, Henderson B, Branca G, Mann W (2011) Climate change mitigation finance for smallholder agriculture. A guide book to harvesting soil carbon sequestration benefits. Food and Agriculture Organization of the United Nations (FAO) report. FAO, Rome

MacDicken KG (1997) A guide to monitoring carbon storage in forestry and agroforestry projects. Winrock International, Little Rock

McBratney AB, Minasny B (2010) Comment on "Determining soil carbon stock changes: simple bulk density corrections fail" [Agric Ecosyst Environ 134 (2009) 251–256]. Agric Ecosyst Environ 136:185–186

McKenzie N, Ryan P, Fogarty P, Wood J (2000) Sampling, measurement and analytical protocols for carbon estimation in soil, litter and coarse woody debris. National Carbon Accounting System Technical Report No. 14, Australian Greenhouse Office, Canberra

Milne E, Neufeldt H, Smalligan M, Rosenstock T, Bernoux M, Bird N, Casarim F, Deneuf K, Easter M, Malin D, Ogle S, Ostwald M, Paustian K, Pearson T, Steglich E (2012) Overview paper: methods for the quantification of net emissions at the landscape level for developing countries in smallholder contexts. CCAFS Report No. 9. CGIAR Research Program on Climate Change, Agriculture and Food Security, Copenhagen

Milne E, Neufeldt H, Rosenstock T, Smalligan M, Cerri CE, Malin D, Easter M, Bernoux M, Ogle S, Casarim F (2013) Methods for the quantification of GHG emissions at the landscape level for developing countries in smallholder contexts. Environ Res Lett 8:015019

Murphy B, Badgery W, Simmons A, Rawson A, Warden E, Andersson K (2013) Soil testing protocols at the paddock scale for contracts and audits – Market-based instrument for soil carbon. New South Wales Department of Primary Industries, Australia

Pearson TRH, Brown S, Ravindranath NH (2005) Integrating carbon benefit estimates into GEF Projects. UNDP, GEF

Prudencio CY (1993) Ring management of soils and crops in the West African semi-arid tropics: the case of the Mossi farming system in Burkina Faso. Agric Ecosyst Environ 47:237–264

Ravindranath NH, Ostwald M (2008) Carbon inventory methods: handbook for Greenhouse Gas Inventory, Carbon Mitigation and Roundwood Production Projects. Advances in Global Change Research 29. Springer, New York

Reeves JB (2010) Near-versus mid-infrared diffuse reflectance spectroscopy for soil analysis emphasizing carbon and laboratory versus on-site analysis: where are we and what needs to be done? Geoderma 158:3–14

Rosenstock TS, Rufino MC, Chirinda N, Bussel L, Reidsma P, Butterbach-Bahl K (2016) Scaling point/plot measurements of greenhouse gas fluxes, balances and intensities to whole-farms and landscapes. In: Rosenstock TS, Rufino MC, Butterbach-Bahl K, Wollenberg E, Richards M (eds) Methods for measuring greenhouse gas balances and evaluating migration options in smallholder agriculture. Springer, New York

Saiz G, Bird M, Domingues T, Schrodt F, Schwarz M, Feldpausch T, Veenendaal E, Djagbletey G, Hien F, Compaore H, Diallo A, Lloyd J (2012) Variation in soil carbon stocks and their determinants across a precipitation gradient in West Africa. Glob Chang Biol 18:1670–1683

Saiz G, Bird M, Wurster C, Quesada CA, Ascough P, Domingues T, Schrodt F, Schwarz M, Feldpausch TR, Veenendaal E, Djagbletey G, Jacobsen G, Hien F, Compaore H, Diallo A, Lloyd J (2015) The influence of C_3 and C_4 vegetation on soil organic matter dynamics in contrasting semi-natural tropical ecosystems. Biogeosciences 12:5041–5059

Shepherd KD, Walsh MG (2002) Development of reflectance libraries for characterization of soil properties. Soil Sci Soc Am J 66:988–998

Shepherd KD, Walsh MG (2007) Infrared spectroscopy—enabling an evidence-based diagnostic surveillance approach to agricultural and environmental management in developing countries. J Near Infrared Spectrosc 15:1–19

Singh K, Murphy BW, Marchant BP (2013) Towards cost-effective estimation of soil carbon stocks at the field scale. Soil Res 50:672–684

Six J, Callewaert S, Lenders S, De Gryze S, Morris SJ, Gregorich EG, Paul EA, Paustian K (2002) Measuring and understanding carbon storage in afforested soils by physical fractionation. Soil Sci Soc Am J 66:1981–1987

Smith P (2004) How long before a change in soil organic carbon can be detected? Glob Chang Biol 10(11):1878–1883

Smith P, Falloon P, Smith JU, Powlson DS (2001) Soil Organic Matter Network (SOMNET): 2001 model and experimental metadata. GCTE Report 7, 2nd ed. GCTE Focus 3 Office, Wallingford

Snedecor GW, Cochran WG (1967) Statistical methods. Iowa State University Press, Ames, p 274

Stevens A, Van Wesemael B, Vandenschrick G, Touré S, Tychon B (2006) Detection of carbon stock change in agricultural soils using spectroscopic techniques. Soil Sci Soc Am J 70:844–850

Subedi BP, Pandey SS, Pandey A, Rana EB, Bhattarai S, Banskota TR, Charmakar S, Tamrakar R (2010) Guidelines for measuring carbon stocks in community managed forests. ANSAB, Kathmandu

Tittonell P, Vanlauwe B, Leffelaar PA, Rowe E, Giller KE (2005a) Exploring diversity in soil fertility management of smallholder farms in western Kenya. I. Heterogeneity at region and farm scale. Agric Ecosyst Environ 110:149–165

Tittonell P, Vanlauwe B, Leffelaar PA, Shepherd KD, Giller KE (2005b) Exploring diversity in soil fertility management of smallholder farms in western Kenya. II. Within-farm variability in resource allocation, nutrient flows and soil fertility status. Agric Ecosyst Environ 110: 166–184

Tittonell P, Muriuki A, Klapwijk CJ, Shepherd KD, Coe R, Vanlauwe B (2013) Soil heterogeneity and soil fertility gradients in smallholder farms of the East African Highlands. Soil Sci Soc Am J 77:525–538

Vågen T-G, Winowiecki LA, Tamene Desta L, Tondoh JE (2015) The Land Degradation Surveillance Framework (LDSF)—Field Guide *v*4.1. World Agroforestry Centre, Nairobi, 14p

Veldkamp A, Kok K, de Koning GHJ, Verburg PH, Priess J, Bergsma AR (2001) The need for multi-scale approaches in spatial specific land use change modelling. Environ Model Assess 6:111–121

Verified Carbon Standard (2012) VCS Module VMD0021: estimation of stocks in the soil carbon pool. VCS, Washington, DC. www.v-c-s.org/methodologies/estimation-stocks-soil-carbon-pool-v10

Wuest SB (2009) Correction of bulk density and sampling method biases using soil mass per unit area. Soil Sci Soc Am J 73:312–316

Wurster CM, Saiz G, Calder A, Bird MI (2010) Recovery of organic matter from mineral-rich sediment and soils for stable isotope analyses using static dense media. Rapid Commun Mass Spectrom 24:165–168

Wynn JG, Bird MI (2007) C_4-derived soil organic carbon decomposes faster than its C_3 counterpart in mixed C_3/C_4 soils. Glob Chang Biol 13:2206–2217

Wynn JG, Bird MI, Vellen L, Grand-Clement E, Carter J, Berry SL (2006) Continental-scale measurement of the soil organic carbon pool with climatic, edaphic, and biotic controls. Glob Biogeochem Cycles 20, GB1007

Chapter 8
Yield Estimation of Food and Non-food Crops in Smallholder Production Systems

Tek B. Sapkota, M.L. Jat, R.K. Jat, P. Kapoor, and Clare Stirling

Abstract Enhancing food security while contributing to mitigate climate change and preserving the natural resource base and vital ecosystem services requires the transition to agricultural production systems that are more productive, use inputs more efficiently, are more resilient to climate variability and emit fewer GHGs into the environment. Therefore, quantification of GHGs from agricultural production systems has been the subject of intensive scientific investigation recently to help researchers, development workers, and policy makers to understand how mitigation can be integrated into policy and practice. However, GHG quantification from smallholder production system should also take into account farm productivity to make such research applicable for smallholder farmers. Therefore, estimation of farm productivity should also be an integral consideration when quantifying smallholder mitigation potential. A wide range of methodologies have been developed to estimate crop yields from smallholder production systems. In this chapter, we present the synthesis of the state-of-the-art of crop yield estimation methods along with their advantages and disadvantages. Besides the plot level measurements and sampling, use of crop models and remote sensing are valuable tools for production estimation but detailed parameterization and validation of such tools are necessary before such tools can be used under smallholder production systems. The decision on which method to be used for a particular situation largely depends on the objective, scale of estimation, and desired level of precision. We emphasize that multiple approaches are needed to optimize the resources and also to have precise estimation at different scales.

T.B. Sapkota (✉) • M.L. Jat • P. Kapoor
International Maize and Wheat Improvement Centre (CIMMYT), G-2, B-Block, National Agricultural Science Centre (NASC) Complex, Dev Prakash Shastri Marg, New Delhi 110 012, India
e-mail: T.Sapkota@cgiar.org

R.K. Jat
Borlaug Institute of South Asia, Samastipur, Pusa, Bihar 848125, India

International Maize and Wheat Improvement Centre (CIMMYT), New Delhi, India

C. Stirling
International Maize and Wheat Improvement Centre (CIMMYT), Wales, UK

T.S. Rosenstock et al. (eds.), *Methods for Measuring Greenhouse Gas Balances and Evaluating Mitigation Options in Smallholder Agriculture*,
DOI 10.1007/978-3-319-29794-1_8

8.1 Introduction

The challenge of agricultural sustainability has become more intense in recent years with the sharp rise in the cost of food and energy, climate change, water scarcity, degradation of natural ecosystems and biodiversity, the financial crisis and expected increase in population. With increasing demands for food and agricultural products, intensification of smallholder production system becomes increasingly necessary. Recently, agricultural technologies that increase food production sustainably while offering climate change adaptation and mitigation benefits–collectively known as climate smart agricultural (CSA) practices-have been the subject of scientific investigation. CSA practices are designed to achieve agricultural sustainability by implementation of sustainable management practices that minimize environmental degradation and conserve resources while maintaining high-yielding profitable systems, and also improve the biological functions of the agroecosystems. However, simultaneous quantification of productive, adaptive, and mitigative production systems is still scant and scattered.

Understanding the greenhouse gas (GHG) fluxes between agricultural fields and the atmosphere is essential to know the contribution of farm practices to GHG emissions. However, quantification of GHGs from agricultural production systems in smallholder systems is meaningless if the livelihood effects of those activities are ignored (Linquist et al. 2012). As farm productivity is inextricably linked to food security of smallholder farmers in developing countries, the importance of productivity must be taken into account in mitigation decision-making and the GHG research agenda supporting those decisions. Most of the GHG emission studies, so far, highlight the emission reduction potential of particular activities without paying due attention on yield and livelihood benefits for smallholder production (Rosenstock et al. 2013). The benefit of smallholder production systems, in terms of reduced emissions and increased carbon sequestration should, therefore, be assessed taking household benefits such as resilience led-productivity enhancement and input use efficiency in due consideration. In this chapter, we focus on comparative analysis of yield estimation methods from field to landscape level under smallholder production practices.

8.2 Crop Productivity Estimation

Various methods have been developed for quantifying production and productivity of agricultural systems at research plot level and also for agricultural statistics at regional and national level. However, as agricultural production systems are changing to address new challenges, for example, CSA practices, the yield estimation methods developed and tested for a particular production system may not adequately reflect the yield for new production systems. For example, the standard crop cut method using sampling frames may create significant bias and error if applied to crops planted in raised beds in row geometry.

Standardization of crop yield estimation methods, particularly in the context of smallholder production system at various scales (field, farm to landscape scale) helps not only to obtain accurate agricultural statistics but also in assessing suitability of low-emission agricultural practices under various production environments. Accurate yield estimation allows trade-off analysis on crop yield and emission reduction of particular production practices thereby helping appropriate mitigation decision-making without compromising smallholder livelihoods and rural development (Rosenstock et al. 2013). This is particularly important in the context that a significant proportion of developing countries have expressed an interest in GHG mitigation in the agriculture sector (Wilkes et al. 2013). Here, we present various yield estimation methods followed by comparative analysis of those methods at various scales i.e., from field to landscape level.

8.2.1 *Crop Cuts*

Estimating crop yield by sampling a small subplot within cultivated field was developed in the 1950s in India (Fermont and Benson 2011) and rapidly adopted as the standard method of crop yield estimation, known popularly as the crop cut method. In this method, yield in one or more subplots is measured and total yield per unit area is calculated as total production divided by total harvested area in the crop cut plot or subplot. The number of subplots and area of each subplot to be selected for yield estimation through crop cuts depends on the resources available and the level of precision required in the estimation. In practice, 1–5 subplots of 0.25–50 m^2 are used for yield estimation. In on-farm research conducted by CIMMYT, use of a 0.5 m by 0.5 m sampling frame overestimated the wheat yield by more than two times as compared to 1 m^2 or larger sampling frame (Fig. 8.1). This finding suggests that when estimating crop yield by using crop cut method, the size of sampling plot should be at least 1 m^2. In the field with variable crop performance, it is advisable to use even larger sampling frame or increase the number of subplots to be harvested for yield estimation. For better result, the person throwing the sampling frame in the field should be blindfold. Alternatively, a person independent of the research or demonstration should throw the sampling frame in the field to minimize the bias.

8.2.2 *Farmers' Survey*

Estimating crop production through farmers' interviews involves asking farmers to estimate or recall the yield for an individual plot, field, or farm. It can be done before harvesting (estimate) or after harvesting (recall). Before harvesting, farmers are asked to predict what quantity they expect to harvest. Farmers will base their predictions of expected yield on previous experiences, by comparing the current crop performance to previous crop performances. Singh (2013) argue that yield estimation surveys following this method should be made at maximum crop growth stage. This helps enumerators/extension worker to verify the farmer's response by visual observation of the crop. Postharvest estimations are commonly made at the

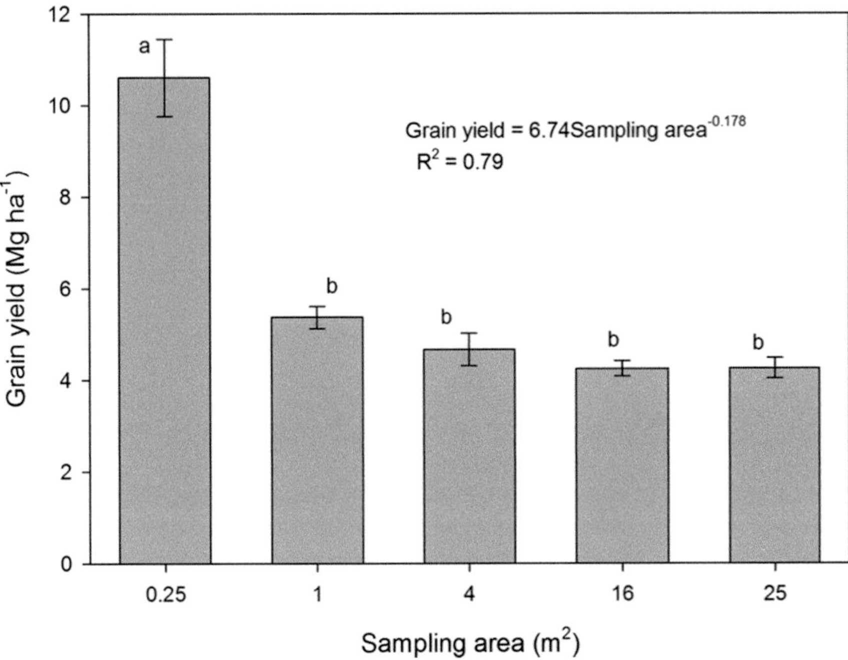

Fig. 8.1 Estimated grain yield of wheat by harvesting the subplot of different size

farmer's house or at the site where the harvest is stored in order for the enumerator to cross-check the estimates with the harvested products. Postharvest surveys should be carried out as soon as farmers harvest the crop, although Erenstein et al. (2007) reported that farmers can recall yield for up to three-to-six previous seasons.

To estimate the crop yield, production data obtained from farmer recall or prediction require division by the plot area from which the crop was or will be harvested. This introduces an additional source of error. To remove this error source, Fermont et al. (2009) obtained a direct estimate of average crop yield by asking farmers to estimate the number of local harvest units they would have obtained from a well-known unit of land, often the farm compound, if it had been planted to a specific crop.

8.2.3 Estimating Crop Yield by Using Grain Weight (Test Weight)

Estimating crop yield by using pre-estimated test weight is one of the easiest and quickest methods which can be used in a number of situations and farm conditions. This is similar to the crop cut method but does not require harvesting and subsequent weighing of the sampled area. Using a sampling frame, count the number of

Table 8.1 Thousand grain weight of some example crops

Crop	1000-grain weight (g)	Source
Wheat	30–45	Jat et al. (2014)
Rice	18–23	Jat et al. (2014)
Lentils	30–50	http://www.depi.vic.gov.au/
Field pea	200	http://www.depi.vic.gov.au/
Chickpea (desi)	180	http://www.depi.vic.gov.au/
Chickpea (kabuli)	380–420	Frade and Valenciano (2005)
Maize	237–268	Sampathkumar et al. (2013)

earheads/pods in 1 m^2 area at least five to seven times within a plot whose yield is to be determined and calculate average number of heads/pods per meter square area. Similarly, count the number of grains in 20–25 heads/pods and take the average. The yield of the crop can then be determined by using the following formula. The 1000-grain weight can be taken from previous data or from published figures (Table 8.1).

$$\text{Yield Mg ha}^{-1} = \frac{\#\,\text{grains per head} \times \#\,\text{heads per m}^2}{100} \times \frac{1000 - \text{grain weight}\,(\text{g})}{1000}$$

The 1000-grain weight of crops is influenced by many factors such as genotype, management, and environment. Therefore, care should be taken to use appropriate 1000-grain weight value based on the variety grown and the growing condition. Estimation accuracy, regardless of the method, depends on the accuracy of observations taken in the field. Counts of grain per head and heads per square meter area must be accurate and taken randomly at enough locations (at least 5) to provide an average of the whole field.

8.2.4 *Whole Plot Harvest*

Harvesting the entire field to determine crop yield is normally done in trial plots, excluding one or more boundary lines that may not reflect the tested treatment due to boundary effects. This method can be employed in experimental or demonstration plots. It can also be used to estimate yield from small-scale farmers' fields if farmers are willing to cooperate but is too costly for larger samples of farmers. The complete harvest method is considered the most accurate and often used as a standard for comparing effectiveness and accuracy of other methods. Crops that have a defined maturity date, such as cereals or legumes with a determinate growth habit, can be harvested in a single operation whereas crops with staggered maturity such as banana, cassava, and legumes or with an indeterminate growth habit like common bean, cowpea, and mungbean require multiple harvests per plot. In many cases, a farmer gathers all his/her produce from his/her land in one place, threshes there and take home the produce after weighing. In such cases, it is easy to estimate the yield by dividing the total yield by the total area cultivated by the farmer.

8.2.5 Sampling for Harvest Unit

This is similar to yield estimation through whole plot harvest except that only a few samples out of the total harvest are weighed. In this method, the number of units, such as sacks, baskets, bundles, is counted after the farmer harvests his/her plot. A number of harvest units are then randomly selected and weighed to obtain an average unit weight. Total harvest of the plot is obtained by multiplying the total number of units harvested by the average unit weight. Crop productivity can then be calculated by dividing total production by the area from where the production came from. Ideally, sampling of harvest units is done just before storage and includes a measurement of the moisture content of the harvested product (Casley and Kumar 1988). This method can be used on larger samples than is possible with crop-cut or whole-plot harvest method. However, the crops must be harvested all at once for this method to be applicable.

An alternative method which requires the physical threshing of only a small sample to estimate yield, biomass, and other yield-related parameters has been developed by Castellanos-Navarrete et al. (2013). This is rather a simple procedure that dramatically reduces the labor and large-scale threshing required to obtain reliable yield and associated yield-related parameters. The methodology can also be used for any situation and any cereal crop. It can be readily applied for on-farm research situations where samples are taken in the field and then transported back to a central point for threshing. Harvest should be done as soon after physiological maturity as possible. Here, after harvesting the crop from sample harvest area, 50–200 tillers are selected randomly for fresh and dry biomass weight, grain weight, and test weight. The yield and yield-related parameters are then determined by using the relationship of the determined parameters and the harvest area. Step-by-step procedures for yield estimation following this method can be found in Castellanos-Navarrete et al. (2013).

8.2.6 Expert Assessment

Sometimes crop yield is estimated by summarizing the opinions of field agronomists, extension agents, and researchers (Dumanski and Onofrei 1989). These experts are often able to estimate crop production or yield by visually assessing the crop condition, such as color, plant vigor, plant density, in the field. This is known as eye assessment. Eye assessment can be combined with field measurement and empirical formulas, collectively known as the expert assessment method. The expert assessment method can be applied on a relatively large scale as compared to the crop-cut method but on a smaller scale than the farmer's estimate. However, eye estimation of crop yield requires not only practical but also technical familiarity with the yield potential of different varieties of crops in different environments. Therefore, accuracy of the yield assessment, in this method, will strongly depend on the level of expertise of the personnel involved in the assessment. Care should be taken not to use extension worker as expert for yield estimation in their own work area as they may be biased to demonstrate their own work (Casley and Kumar 1988).

8.2.7 *Crop Cards*

The crop card method is a refined version of the farmer recall procedure to obtain more reliable harvest estimates for crops with an extended harvest period or multiple harvests, such as cassava, banana, cowpea, sweet potato. As farmers may have problems in accurately remembering the amounts they harvested over time from one or several plots, this method helps them by keeping the written record of all plots. In this method, each farmer in a survey is given a set of crop cards where he/she records the quantity of crop in each harvest, which can then be added up to calculate the total harvested yield. However, this may be challenging to use in smallholder production contexts of developing countries due to high illiteracy rates and lack of adequate manpower for regular monitoring (Ssekiboobo 2007).

8.2.8 *Crop Modelling*

Crop modelling is widely used to estimate average biological yields in the conditions of smallholder farmers. Empirical–statistical crop models establish a relationship between yield and environmental factors from long-term datasets and use the established relationship to predict crop yield at regional or national levels based on environmental data (Park et al. 2005). Empirical crop growth models are relatively simple to develop, but these models cannot take into account the temporal changes in crop yields without long-term field experiments (Jame and Cutforth 1996). Furthermore, the derived functional equation is locally specific, and it is thus difficult to extrapolate to other areas unless environmental conditions are similar. Many of such models embody a number of simplifications. For example, weeds, diseases, and insect pests are assumed to be controlled, and there are no extreme weather events such as heavy storms.

Process-based crop models, on the other hand, estimate crop yield on the basis of daily gains in biomass production by taking into account all known interactions between physiological processes and environmental conditions (Sawasawa 2003). Because process-based models explicitly include plant physiology, agroclimatic conditions, and biochemical processes, these models are able to simulate both temporal and spatial dynamics of crop yields and thus have higher extrapolation potential than empirical models.

8.2.9 *Allometric Models*

Allometric models are mathematical relationships between plant morphological characteristics and crop yield. The morphological characters can be measured on a selected number of plants which then can be used to predict biological yield in field. Allometric models should be based on variables that can be quantified easily using rapid, inexpensive, and non-destructive methods of data collection (Fermont and Benson 2011). For bananas in Uganda, Wairegi et al. (2009) found that a multivariate model using girth of the pseudo-stem at base and at 1 m, the number of hands, and the number of fingers gave a robust prediction of bunch weight. Tittonell et al. (2005) used plant height and ear length to predict maize yields in western Kenya. In cereal crops, the number of tillers per unit area, ear or spike length, number of grains per spike, and 1000-grain weight—commonly known as yield attributing characters—can be determined and used to estimate the crop yield. Data collection is one of the prerequisites of this method although data collection may be less labor intensive than with the crop cut method.

8.2.10 *Remote Sensing*

Use of remote sensing to estimate the biological crop yield is being explored in many countries and likely will become the basis of agricultural statistics in the future (Zhao et al. 2007). Crop yield estimation using remote sensing is based on the principle of spectral reflectance of green plants, which can be captured in satellite images as spectral data, and depends on the state, structure, and composition of the plant. The spectral data can be used to construct several vegetation indices such as normalized difference vegetation index (NDVI) which indicates the green biomass that can be used as proxy indicator of the yield (Prasad et al. 2006). The limitation in the use of satellite images to estimate crop yields of smallholder farmers is that the resolution of available satellite imagery (pixel size) is not sufficiently detailed to capture the variability of crops and crop performance in smallholder fields, which often are less than 0.1 ha in size and sometimes intercropped (Fermont and Benson 2011). In India, for example, vegetation indices from satellite images show only a moderate correlation (R^2 between 0.45 and 0.54) with crop cut data (Singh 2013).

8.3 Critical Analysis and Comparison of Yield Estimation Methods with Regards to Cost, Scale, and Accuracy

A comparison of the wide range of methodologies to estimate crop production in terms of their cost effectiveness, suitability for different scales from field to landscape and sources of errors or bias is presented in Table 8.2. A strong advantage of

Table 8.2 Comparison of various methods of crop production estimation with regard to their cost-effectiveness, scale, and accuracy

Method	Cost-effectiveness	Scale	Precision in estimation, errors, and biases
Crop cut	Time and labor intensive	Field, farm, and sometimes landscape level	Tendency to overestimate
Farmer's estimate	Cheap and quick method that saves time and money	Farm to landscape	Fairly accurate estimation but needs adequate supervision. Subjective. Sometimes farmers deliberately overestimate or underestimate
Sampling harvest unit	Cost-effective	Farm to landscape	Error prone in the condition where farmers harvest from multiple areas at time and not possible with staggered harvesting
Whole plot harvest	Cost intensive, labor intensive	Plot level, farm level, case study	Almost bias/error-free
Expert assessment	Moderately cost-effective	From farm to landscape level	Chances of error increases if different teams of experts are used or extension people are used to estimate yield in their own area. Subjective
Crop cards	Cost and labor intensive	Field to farm level	Bias due to illiteracy, use of local units etc.
Crop modelling	Cost-effective	Landscape	Less if adequately parameterized and calibrated. Do not include induced improvements in agricultural technology
Purchaser's/ insurance record	Cost-effective	Field scale	Suitable for cash crops only with no household consumption
Allometric models	Cost-effective	Field scale	Suitable for few crops only
Remote sensing	Cost-effective	Landscape	Chance of error in cases where different crops have the same signature

the crop-cut method is that the area of the cut is known and thus does not introduce an error into the final yield computation. It has been a standard method for yield estimation recommended by organizations such as the Food and Agriculture Organization of the United Nations for years. However, crop cuttings may suffer from serious limitation due to heterogeneity of crop conditions within farmers' plots. In crop cuts, enumerators have the tendency not to sample locations with poor crop stand, leave border areas where crop yield is generally lower than in the middle of the plot and include the plant falling at the edge of sampling frame. A study done in Bangladesh found that even with best-educated enumerators, crop-cut estimates exceeded actual yield by 20 % whereas farmers' estimates of production were lower (Diskin 1999). Further, crop cut only estimates biological yield without taking into account postharvest losses and is therefore unable to estimate economic yield, which is of most interest to farmers. All these tendencies contribute to upward bias when extrapolating results to a larger area. Further, using a large weighing balance to weigh smaller quantities from crop cuts may sometimes introduce measurement errors. This method is costly and time-consuming, and not suitable for heterogeneous crop performance (typical of smallholder production systems) and staggered harvesting as this is a one-point-in-time measurement.

The farmers' estimation method does not require laborious measurements, and therefore this method is time- and cost-efficient and is suitable for estimation at larger scales. For years, it was assumed that farmers' estimates were too subjective and unreliable and when differences appeared between crop cut and farmers' production estimates, it was attributed as farmers' error. However, research in 1980s suggested that farmers' estimation may be just as accurate as crop cut, at least for determining total farm production (Murphy et al. 1991). However, literacy levels of farmers and non-standard harvest units pose serious drawbacks in its use in smallholder production systems of developing countries. Farmers may use part of their produce as in-kind payment to their labor which they may not count in their estimate, leading to underestimation. Further, many farmers consciously over- or underestimate in the case of suspected benefits such as food aid or penalties such as taxes (Diskin 1999). Expert assessment can be relatively error-free if the same team of experts can be used throughout the study (Rozelle 1991). However, finding a large number of experts with required practical and technical experience to estimate relative performance of different crops/varieties under different environments is a challenge to employ this approach at larger scales. Furthermore, both farmer's estimation and expert assessment are subjective and amenable to several non-sampling errors. Therefore, it is advisable to combine these methods with other methods for better estimation of crop yield.

The advantage of whole plot harvest method is that it is almost bias-free since all sources of possible errors and biases associated with crop cut or farmers' estimate are eliminated when the entire field is harvested. However, this involves a large volume of work to obtain robust estimates of yield at landscape level. Sampling of harvest units can be used on larger samples than is possible with crop-cut or whole plot harvest method. However, this method is unsuitable for crops with staggered harvesting.

Use of crop cards can be combined with farmers' estimate for crops with multiple harvesting and staggered ripening. However, this is again very labor intensive and cannot be employed for large-scale surveys. Further, use of local unit of measurement by different farmers may introduce error in estimation. Use of allometric methods is limited to a certain number of crops such as banana and maize. In developed countries, purchasers' records or crop insurance data may be used for crop yield estimation but this method may not be suitable in the context of smallholder production in developing countries.

Crop modelling and remote sensing are cost-effective methods of yield estimation which can be employed at large scales fairly accurately although empirical models fail to capture landscape heterogeneity and process-based models need rigorous parameterization, calibration, and validation before they can be used for large-scale estimation.

8.4 Conclusion

Precise estimation of crop yield in smallholder agriculture is challenging because of highly heterogeneous crop performance within a plot, continuous planting and intercropping or mixed cropping to meet various household requirements. Staggered ripening of many crops with an extended harvest period and planted area not being equal to harvested area further complicates the issue of crop yield determination in smallholder farmers' condition. A wide range of methodologies have been developed to estimate crop yields in the smallholder production systems, each with advantages and disadvantages. This review has primarily considered the application of these methodologies to cereal cropping systems, but the methodologies can be adapted to other cropping systems as well. A choice of method depends on the objective and desired level of precision, scale of estimation, and available resources. For example, whole plot harvesting may be suitable for small-scale detailed studies at plot level whereas for large-scale survey at regional level combination of crop cut, farmer's estimation and expert assessment may be used. Use of crop models and remote sensing may be appropriate for agricultural statistics, provided adequate parameterization of models is done and imagery at sufficiently fine resolution to capture the variability of crops and their performance in smallholder fields is available.

References

Casley DJ, Kumar K (1988) The collection, analysis and use of monitoring and evaluation data. Johns Hopkins University Press for the World Bank, Baltimore

Castellanos-Navarrete A, Chocobar A, Cox RA, Fonteyne S, Govaerts B, Jespers N, Kiennle F, Sayer KD, Verhulst N (2013) Yield and yield components: a practical guide for comparing crop management practices. International Maize and Wheat Improvement Center (CIMMYT). http://repository.cimmyt.org/xmlui/handle/10883/3387. Accessed 29 Jan 2015

Diskin P (1999) Agricultural productivity indicators measurement guide. Food and Nutrition Technical Assistance Project (FANTA), Academy for Educational Development. http://www.fsnnetwork.org/sites/default/files/agric_productivity_indicators.pdf. Accessed 29 Jan 2015

Dumanski J, Onofrei C (1989) Techniques of crop yield assessment for agricultural land evaluation. Soil Use Manage 5:9–15

Erenstein O, Malik RK, Singh S (2007) Adoption and Impact of zero-tillage in the rice-wheat zone of irrigated Haryana, India. International Maize and Wheat Improvement Centre (CIMMYT), New Delhi

Fermont A, Benson T (2011) Estimating yield of food crops grown by smallholder farmers: a review in the Uganda context. International Food Policy Research Institute (IFPRI). http://www.ifpri.org/sites/default/files/publications/ifpridp01097.pdf. Accessed 29 Jan 2015

Fermont AM, van Asten JA, Tittonell P, Van Wijk MT, Giller KE (2009) Closing the cassava yield gap: an analysis from small-holder farms in East Africa. Field Crop Res 112:24–36

Frade MMM, Valenciano JB (2005) Effect of sowing density on the yield and yield components of spring-sown irrigated chickpea (Cicer arietinum) grown in Spain. New Zeal J Crop Hort Sci 33:367–371

Jame YW, Cutforth HW (1996) Crop growth models for decision support systems. Can J Plant Sci 76:9–19

Jat RK, Sapkota TB, Singh RG, Jat ML, Kumar M, Gupta RK (2014) Seven years of conservation agriculture in a rice–wheat rotation of Eastern Gangetic Plains of South Asia: yield trends and economic profitability. Field Crop Res 164:199–210

Linquist B, Groenigen KJ, Adviento-Borbe MA, Pittelkow C, Kessel C (2012) An agronomic assessment of greenhouse gas emissions from major cereal crops. Glob Chang Biol 18:194–209

Murphy J, Casley DJ, Curry JJ (1991) Farmers' estimations as a source of production data. World Bank technical paper no. 32, p 80

Park SJ, Hwang CS, Vlek PLG (2005) Comparison of adaptive techniques to predict crop yield response under varying soil and land management conditions. Agric Syst 85:59–81

Prasad AK, Chai L, Singh RP, Kafatos M (2006) Crop yield estimation model for Iowa using remote sensing and surface parameters. Int J Appl Earth Obs Geoinf 8:26–33

Rosenstock TS, Rufino MC, Butterbach-Bahl K, Wollenberg E (2013) Toward a protocol for quantifying the greenhouse gas balance and identifying mitigation options in smallholder farming systems. Environ Res Lett 8:021003

Rozelle S (1991) Rural household data collection in developing countries: designing instruments and methods for collecting farm production data. Cornell University working paper in agricultural economics 91(17)

Sampathkumar T, Pandian BJ, Rangaswamy MV, Manickasundaram P, Jeyakumar P (2013) Influence of deficit irrigation on growth, yield and yield parameters of cotton–maize cropping sequence. Agric Water Manage 130:90–102

Sawasawa HLA (2003) Crop yield estimation: integrating RS, GIS and management factor. A case study Birkoor Kortigiri Mandals, Nizamabad District India. Thesis, International Institute for Geo-Information Science and Earth Observation

Singh R (2013) Use of satellite data and farmers eye estimate for crop yield modeling. J Indian Soc Agric Stat 56(2):166–176

Ssekiboobo AM (2007) Practical problems in the estimation of performance indicators for the agricultural sector in Uganda. In: Fourth international conference on agricultural statistics, October 22–24, Beijing, China

Tittonell P, Vanlauwe B, Leffelaar P, Giller KE (2005) Estimating yields of tropical maize genotypes from non-destructive, on-farm plant morphological measurements. Agric Ecosyst Environ 105:213–220

Wairegi LWI, Van Asten PJA, Tenywa M, Bekunda M (2009) Quantifying bunch weights of the East African Highland bananas (Musa spp. AAA-EA) using non-destructive field observations. Sci Hortic 121:63–72

Wilkes A, Tennigkeit T, Solymosi K (2013) National planning for GHG mitigation in agriculture: a guidance document. Food and Agricultural Organization of the United Nations. http://www.fao.org/docrep/018/i3324e/i3324e.pdf. Accessed 29 Jan 2015

Zhao J, Shi K, Wei F (2007) Research and application of remote sensing techniques in Chinese agricultural statistics. In: Fourth international conference on agricultural statistics, October 22–24, Beijing, China

Chapter 9
Scaling Point and Plot Measurements of Greenhouse Gas Fluxes, Balances, and Intensities to Whole Farms and Landscapes

Todd S. Rosenstock, Mariana C. Rufino, Ngonidzashe Chirinda, Lenny van Bussel, Pytrik Reidsma, and Klaus Butterbach-Bahl

Abstract Measurements of nutrient stocks and greenhouse gas (GHG) fluxes are typically collected at very local scales (<1 to 30 m^2) and then extrapolated to estimate impacts at larger spatial extents (farms, landscapes, or even countries). Translating point measurements to higher levels of aggregation is called *scaling*. Scaling fundamentally involves conversion of data through integration or interpolation and/or simplifying or nesting models. Model and data manipulation techniques to scale estimates are referred to as scaling methods.

In this chapter, we first discuss the necessity and underlying premise of scaling and scaling methods. Almost all cases of agricultural GHG emissions and carbon (C) stock change research relies on disaggregated data, either spatially or by farming activity, as a fundamental input of scaling. Therefore, we then assess the utility of using empirical and process-based models with disaggregated data, specifically concentrating on the opportunities and challenges for their application to diverse smallholder farming systems in tropical regions. We describe key advancements needed to improve the confidence in results from these scaling methods in the future.

T.S. Rosenstock (✉)
World Agroforestry Centre (ICRAF), PO Box 30677-00100, UN Avenue-Gigiri, Nairobi, Kenya
e-mail: t.rosenstock@cgiar.org

M.C. Rufino
Center for International Forestry Research (CIFOR), Nairobi, Kenya

N. Chirinda
International Center for Tropical Agriculture (CIAT), Cali, Colombia

L. van Bussel • P. Reidsma
Wageningen University and Research Centre, Wageningen, Netherlands

K. Butterbach-Bahl
International Livestock Research Institute (ILRI), Nairobi, Kenya

Karlsruhe Institute of Technology, Institute of Meteorology and Climate Research, Atmospheric Environmental Research (IMK-IFU), Kreuzeckbahnstr. 19, Garmisch-Partenkirchen, Germany

T.S. Rosenstock et al. (eds.), *Methods for Measuring Greenhouse Gas Balances and Evaluating Mitigation Options in Smallholder Agriculture*,
DOI 10.1007/978-3-319-29794-1_9

9.1 Introduction?

Agricultural systems are a major source of atmospheric greenhouse gas (GHG) emissions, contributing approximately 30 % to total anthropogenic emissions if land use change is included (Vermeulen et al. 2012). To better target interventions aimed at reducing GHG emissions from agricultural systems, there is a need for information on GHG balances and the GHG intensity of agricultural products (e.g., emissions per unit product) at levels where livelihood and environmental impacts occur and land management decisions are being made. However, even in smallholder farming systems where decisions are taken on fields and farms that are usually less than 1 ha, this decision scale is substantially greater than the scale at which changes in GHG fluxes take place or are measured, often that of micrometers and meters (Butterbach-Bahl et al. 2013). The factors regulating nitrous oxide (N_2O) generation in agricultural fields illustrate this point. At the scale of soil aggregates—mm in size–soil moisture affects oxygen available to microbes, driving denitrification (the conversion of NO_3^- to N_2O principally by facultative anaerobic bacteria). Meanwhile, soil moisture, influenced by the percentage of water filled pore space, is regulated by precipitation and soil tillage—events determined at a larger spatial extent. Furthermore, heterogeneous distribution of decomposing residues from the previous harvest may lead to formation of denitrification and N_2O hotspots at the cm scale, thereby triggering changes in the magnitude and spatial variability of fluxes even at plot scale (Groffman et al. 2009). Consequently, land-based mitigation actions require a lower resolution of information than that needed to explain the processes driving GHG emissions at the soil–plant–atmosphere interface.

GHG fluxes are typically measured at locations or "points," intended to be representative of a larger area. Independent of source, sink or molecule, GHG measurements—for example chamber measurements of fluxes—are conducted on only a fraction of the area or a few of the landscape units because of costs and complexity (Rufino et al. 2016; Butterbach-Bahl et al. 2016). When attempting to understand landscape or regional GHG fluxes or consider mitigation options, it is therefore necessary that these point measurements be translated to larger extents where effective and meaningful mitigation actions can be taken.

"Scaling" GHG flux measurements underlies GHG accounting (e.g., national inventories), and forms the basis for policy analysis (e.g., marginal abatement cost curves), development strategies (e.g., low emission development), and even simple testing of mitigation options (e.g., comparing impacts of one practice versus an alternative). Thus, it is important to understand basic principles and terminology that pertain to scales and scaling, to avoid confusion in discussions and analysis. *Scale* refers to the spatial or temporal dimension of a phenomenon (van Delden et al. 2011; Ewert 2004). *Scaling* refers to the transfer of information between scales or organizational levels (Blöschl and Sivapalan 1995). *Scaling methods* refer to tools required to accomplish scaling. This chapter is concerned with understanding the theory and practice behind scaling methods as applied to GHG measurements and impacts.

9.2 Scaling Methods

Most scaling methods are grounded in ecological hierarchy theory. Hierarchy theory provides a conceptual framing for scaling in that it structures systems as nested levels of organization (Holling 1992). Components are arranged within higher levels; for example, a field is part of a farm that can be thought of as part of a landscape; moreover, these different components are spatially heterogeneous areas of interacting patches of ecosystems (Fig. 9.1). Scaling methods rely on this conceptual framing to infer relationships between attributes and to translate values derived from point measurements into estimates across scales.

Scaling methods can be categorized into two groups: (1) manipulation of input or output data or (2) manipulation of models (Volk and Ewert 2011). Approaches that manipulate data are extrapolation, interpolation, (dis)aggregation, or averaging sampled input data (i.e., point measurements) to generate estimates at larger scales (Table 9.1). National Greenhouse Gas Inventories that use IPCC default Tier 1 emission factors (IPCC 2006) are an example of a scaling method that uses a data manipulation approach, namely disaggregation and aggregation. Agriculture is disaggregated into farming activities and their extents (e.g., size of cattle population or tons of nitrogen (N) fertilizer applied) for which a coefficient or empirical model derived from point measurements of the relationship between that activity and GHG fluxes (i.e., empirical model) is then used to calculate emissions at national or subnational levels. Data manipulation approaches are among the simplest approaches to implement, especially in regions and for production conditions where data are sparse. However, since data manipulation approaches generally neglect heterogeneity in GHG emissions and underlying physicochemical and biological processes, estimates may not represent observed fluxes well at the site level. However, in most cases for developing countries, the accuracy of using such methods is unknown because there are insufficient data to evaluate the variation of source events (input data) or the accuracy of outputs. The ability to generate accurate estimates at larger temporal or spatial scales by manipulating data depends on (1) representative sampling of the disaggregated GHG source/sink activities and (2) the availability of a

Fig. 9.1 Illustration of a nested hierarchy. Regions (East Africa) can be disaggregated to landscapes (natural forest, communal lands, and agriculture) to farms (mixed crop–livestock) to fields (cabbages) (Photos: Authors; CCAFS; Google Maps 2015)

Table 9.1 Conceptual framework of select scaling methods based on Ewert et al. (2011). Reprinted with permission.

Scaling method	Graphical representation	Opportunities	Challenges	GHG example
Manipulation of data				
Extrapolation and singling out	Extrapolation / Singling out	Simple	Heterogeneity in inputs are neglected	Tully et al. (in prep)
Aggregation and disaggregation	Aggregation / Disaggregation	Spatial heterogeneity is taken into account	Need to have hypotheses about underlying drivers of input data heterogeneity	Rufino et al. (2016)
Aggregation/ averaging (stratified input data)	Model	Less computationally intensive because of averaged input data	Averaging input data may compromise modeling efforts	Bryan et al. (2013), Li et al. (2005)
Aggregation/ averaging (stratified output data)	Model	More accurate representation of heterogeneity	Data and simulation intensive which limits applicability at scale	De Gryze et al. (2010)
Manipulation of models				
Modification of model parameters	Model / Parameter	Uses existing models	Fine-scale model parameters may be inappropriate for larger scales	
Simplification of model structure	Model / Summary model	Relies on understanding of known fundamental relationships	Subject to availability of data and understanding of processes	Perlman et al. (2013), Spencer et al. (2011)
Derivation of response function or coefficients	Model / Model / Responses	Simplifies process-based model output to summary function	Simplifying relationships may neglect important dynamics.	Sieber et al. (2013)

Based on Ewert et al. (2011)

reasonable model—empirical or process-based—to scale input data. Recently, novel approaches for disaggregation of national, landscape, or farm components such as stratification by socioecological niches using a combination of household surveys and remote sensing and stratification by agroecological conditions using existing climate, soils, and management information have been evaluated to improve estimates because of the better representation of the heterogeneity found in plots, fields, farms, and landscapes (Hickman et al. 2015; Rufino et al. 2016).

The alternative to manipulating data is to modify existing models to be relevant at larger spatial scales. This has been successfully done for national-scale soil C monitoring in the United States, where an integrated data collection and biogeochemical process-based model (DAYCENT) estimates changes in soil C stocks (Spencer et al. 2011). However, other examples for agricultural GHG impact assessments remain scientific exercises (see Perlman et al. 2013 for national scale N_2O assessment). Approaches to manipulate models change the model structure to account for the availability and resolution of input data and to make them computationally tractable. Reformulation of model structure (not creating new models) can result in a reduction of parameters (e.g., macroecological models of functional traits) or simplified model functional forms (e.g., empirical equations derived from multiple runs of process-based models). An important consideration is that scaling by modifying models introduces uncertainty: uncertainty in the quality and quantity of input data, uncertainty of datasets used to test models, and uncertainty related to model structure and parameters in the revised models.

Theory supporting the manipulation of data and models as well as potential errors/uncertainties in outcomes is reviewed in the integrated assessment literature (e.g., Ewert et al. 2011; Volk and Ewert 2011). The process of selecting representative sampling points by various stratification methods (e.g., spatially, land cover, farming activity, etc.) are covered in Chap. 2 and measurement techniques for various fluxes and productivity are covered in Chaps. 3–8. Here we discuss the two methods most commonly used to *scale up* point measurements of disaggregation/aggregation data: empirical and process-based models.

Empirical models are usually relatively simple statistical functions constructed based on the relationship between occurrence of activities or external events, farming or rainfall for example, and monitored responses in the magnitude and temporal and spatial variability of GHG fluxes. By contrast, process-based ecosystem models are built upon our current theoretical understanding of the physicochemical and biological processes underlying GHG emissions. They represent current understanding of complex processes and the interactions of C, N, and water cycling at the ecosystem scale to simulate the mechanisms that control GHG fluxes. However, process models need detailed input information and have numerous parameters describing key ecosystem processes and some of the algorithms are still empirical and represent apparent flux responses rather than the underlying processes. Unlike empirical models that require calibration each time they are used, one assumes that the simulated processes are universal and, thus, that are based on a number of site tests, they might be applied at sites with a different agro-ecological regime for which they have not previously been calibrated, although calibration of specific parameters might still be required. In the following, we briefly describe these two approaches, their applicability for smallholder systems, representation of the landscape units, technical demands of the process, and sources of uncertainty.

9.3 Using Empirical and Process-Based Models with Disaggregated Data

9.3.1 *Empirical Models*

Empirical models for scaling GHGs are based on statistical functions that relate land management "activities" such as extent of a land cover type, amount of fertilizer applied, or the number of heads of livestock to changes in GHG emissions or C sequestration. Carbon stock changes, and GHG fluxes can then be calculated based on two types of input data: (1) that describes the occurrence of activities (the so-called "activity data") and (2) the average effect that an activity has on a nutrient stock or flux in question ("emission factors") (Eq. (9.1)).

$$\mathrm{GHG} = \sum_{i}^{n} A_i * \mathrm{EF}_i \tag{9.1}$$

where

GHG equals the stock (mass) or flux (rate: mass per unit time), sequestration or balance in units of C, N, or an integration of the two (CO_2 eq)
A represents the extent (area) over which an activity occurs
EF is an emissions factor (e.g., a constant rate relative to the specific activity: mass per unit time per unit area)

Summation of GHG fluxes or stock changes across *N* activities (sources/sinks) generates a cumulative balance for the selected area. This approach is analogous to a linear aggregation scaling method based on measurements or estimated values.

The most widely applied empirical models for scaling GHGs are contained within the IPCC Guidelines for Greenhouse Gas Accounting (IPCC 2006). The IPCC Guidelines define global (Tier 1) and, sometimes regional (Tier 2) emission factors for GHG sources and sinks such as the methane produced by enteric fermentation per head of cattle or the amount of nitrous oxide resulting from the application of nitrogenous fertilizers. Persons interested in GHG quantification can multiply these values and use the provided equations with locally relevant data on farm and landscape management activities to generate estimates of individual sources and sinks or cumulative GHG balances. Application of emission factors and empirical models is the foundation of national GHG inventories and data (Tubiello et al. 2013) and is becoming more common for landscape GHG accounting including *ex-ante* climate change mitigation project impact assessments (Colomb and Bockel 2013).

IPCC Tier 1 default emission factors are based on both empirical data and expert opinion. In some cases, emissions factors are derived from analysis of 100 s or even 1000 s of measurements of the source activity and the rates of emissions. For instance, IPCC default emissions factor for nitrous oxide emissions from N fertilizer use (%) are based on the database of nearly 2 000 individual measurements from studies conducted around the world (Stehfest and Bouwman 2006). Distribution of the studies they are taken from is however biased toward measurement campaigns conducted in Europe and North America. Other emission factors are estimated based on very limited data (e.g., single values for carbon stocks in agroforestry systems) or expert opinion (e.g., emission factor for methane emission from enteric fermentation is based on modeled results, not measurements, for Africa) (IPCC 2006). Global default emission factors are published in the National Guidelines for Inventories while other regionally relevant emission factors are available in the IPCC Emissions Factor database, peer-reviewed literature and in the future will be made available through the SAMPLES web platform.

Empirical models are typically thought to generate reasonable approximations of GHG fluxes at higher levels of organizations and large spatial extent (Del Grosso et al. 2008), presuming the activity data are well constrained. This is because it is thought that at large scales such as across countries, the departure of actual fluxes from average emissions factor values will average out with aggregation of multiple land units. However, for any local scale—farms for example, where local environmental and management heterogeneity of conditions are not well represented in the global datasets, applying empirical models and emissions factors may represent a significant departure from actual fluxes.

The relevance of using empirical models for farm-scale estimates of GHG balances is untested and perhaps spurious, especially for farming systems in developing countries. IPCC guidelines using Tier 1 default factors were not designed for this purpose. Tier 1 approaches were intended to be used when the source activity was relatively inconsequential to total GHG budgets, perhaps contributing less than 5 % of the total (IPCC 2006). Furthermore, significant variations in GHG flux rates occur between point locations due to edaphic mechanisms that control biological emission processes. Because observations of GHG fluxes for tropical smallholder farming systems are scarce or nearly missing in available databases, Tier 1 default factors may considerably misrepresent flux rates for such systems. In view of the low use of N fertilizers in sub-Saharan Africa it is therefore not surprising that many of the N_2O fluxes currently being measured there are 1/3 to 1/2 of those estimated using the Tier 1 IPCC emission factors (Hickman et al. 2014; Shcherbak et al. 2014). A comprehensive evaluation of Tier 1 emission factors relating to GHG impacts measured in tropical regions is currently lacking. Despite these concerns and the uncertainty of the results, disaggregation of whole farms into component activities and applying available empirical models remains a way to estimate relative impacts of smallholder farming activities at the whole-farm level (Seebauer 2014), as well as understand emission hotspots and the research gaps.

Emissions from livestock production in the tropics, namely from enteric fermentation and manure management, present their own challenges due to data scarcity (Goopy et al. 2016). Similarly to soil fluxes, emissions from both sources are poorly constrained and according to the review by Owen and Silver (2015) data for dairy manure management are limited in Africa and extremely scarce for other systems (Predotova et al. 2010). Yet in many countries, these sources are thought to be substantial contributors to total GHG budgets (Gerber et al. 2013).

Besides poorly constrained emission factors, an additional issue (and arguably most important) is limited knowledge of farm management practices (*A* in Eq. (9.1)), which limits the use of empirical relationships and models to calculate fluxes. Many developing countries have poorly defined record keeping and reporting schemes about organic and inorganic fertilizer use, manure management, crop rotations, and other activities, so there is limited information on the extent of land management decisions (Ogle et al. 2013). This adds another source of uncertainty (in addition to emission factors themselves). Valentini et al. (2014) reported that estimates of the extent of various land cover types in Africa can be from 2.5 to 110 % different, depending on the data source, either using inventory sources or satellite imagery. Other evidence from data collection methods suggests that the uncertainty in farm management practices is similar to that of emissions, 30–80 % (Fig. 9.3, Seebauer 2014). New practices have been developed to help developing countries better represent the activities in their agricultural landscapes (Tubiello et al. 2013) and many institutions such as the US Environmental Protection Agency train government personnel in developing countries to co-compile inventories. However, problems with the data quality itself remain. Incentives to improve and standardize data collection and archiving efforts are limited.

Simplicity and transparency are the largest benefits of using data (dis)aggregation techniques and empirical models for scaling GHG estimates. The models represent relationships that are easy to understand and implement, which makes them accessible to next users without requiring much technical expertise. This has led to the creation of a wide range of GHG calculators such as the Cool Farm Tool and EX-ACT (see Colomb and Bockel 2013 for a review). These tools make it possible for non-specialists to perform calculations and generate estimates of GHG balances with relatively little data or effort. It is still unknown, however, whether the estimates produced by such tools provide robust values—either in terms of absolute or relative changes between different practices (Fig. 9.2).

9.3.2 Process-Based Models

Empirical models are only one way to scale measured data. Process-based models are also used. For example, Bryan et al. (2013) averaged household data for seven counties and four agroecological zones in Kenya used a process-based model to predict changes in methane emissions from enteric fermentation and revenue with improved feeding practices (Table 9.2). Process-based models consist of equations implementing current scientific understanding of the mechanisms determining system properties. Even though microbial and physicochemical processes involved in GHG emissions from soils are implemented in various biogeochemical models, equations are often based on empirical observations or represent apparent changes in production rates or microbial activity due to, for example, changes in environmental conditions such as changes in moisture and temperature. Thus, models describe a system consisting of components such as soil physics and energy fluxes,

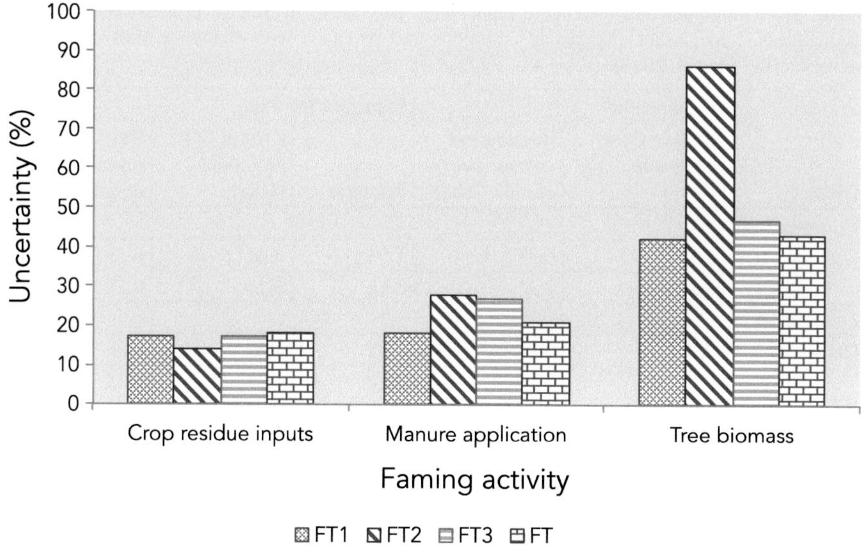

Fig. 9.2 Uncertainty of activity data inputs into a whole-farm accounting approach used in Western Kenya (Seebauer 2014). Uncertainty depends on the farm activity in question and ranges from 10 to 20 % for crop residue inputs up to greater than 80 % with on-farm tree biomass. Data were collected by survey and colors represent different farm types

vegetation biomass development, or soil microbial C and N turnover and their interactions, which are represented by the equations describing states and rates at different points in time (temporal resolution). Process-based GHG models are designed to run at source scale (e.g., site or animal) after being calibrated based on observed relationships in controlled experiments and monitoring data. Because the equations represent principal microbial, biogeochemical and physicochemical processes underlying ecosystem–atmosphere exchange processes and the emission of GHGs, the models can be suitable to simulate GHG dynamics under diverse environmental and management conditions, even conducting "what if" scenario type of experiments. The robustness of process-based models has made them a widely used predictive tool in global change studies and they might be suitable as well to account for fine scale heterogeneity in the farming context, which is not possible with current empirical models. However, process-based models need to be tested for their ability to represent GHG under specific conditions to have confidence in their predictions. This is an involved process, which restricts their utility for sites and systems outside the range of the available calibration data. Until process-based models have been adapted, calibrated, and evaluated to account for diversity and complexity characteristic of smallholder farming, their use for GHG quantification at the whole-farm level in mixed systems, such as the crop–livestock systems of Africa, remains a challenge, requires a tight coupling of sectorial models and a whole system understanding, and implies significant uncertainty.

Table 9.2 Geographically averaged input data was used to run a process-based model (RUMINANT) to predict changes in emissions and revenues with changing diets under two scenarios (Bryan et al. 2013)

	Baseline diet		Improved feeding		
District	Cost of CO_2e emissions (US$)	Baseline net revenue per L of milk (US$)	Scenario	Cost of CO_2e emissions (US$)	Baseline net revenue per L of milk (US$)
Prosopis					
Garissa	6.53	0.33	1.5 kg	6.45	0.23
	6.53	0.33	3 kg	6.16	0.18
Desmodium					
Gem	7.77	0.11	1 kg	7.52	0.26
			2 kg	7.85	0.23
Napier grass					
Mbeere	9.64	0.04	2 kg	9.94	0.16
	9.64	0.04	3 kg	9.90	0.15
Hay					
Othaya	9.57	0.15	2 kg	9.68	0.16
	9.57	0.15	4 kg	9.61	0.11
Grevillia					
Njoro	9.06	0.14	1 kg	9.61	0.19
	9.06	0.14	2 kg	10.63	0.19

The accuracy of a process-based model is related to errors due to model structure (model parameter uncertainty) or errors due to the accuracy of data inputs (input uncertainty). Errors related to model structure are based on incomplete understanding and knowledge of the fundamental relationships that are driving GHG production and consumption processes in soils, variation in ways to describe underlying processes, and fluxes at the soil–atmosphere interface and the representation of them in the model. These errors can be quantified statistically by comparing the model's predicted GHG fluxes to measured GHG fluxes, with correlation coefficients for instance. Errors related to input uncertainty occur because the input data describing a particular system is not well known. This may be particularly problematic in developing countries when the detailed climate, soils, and land use data are not available at a high degree of resolution. Input uncertainty can be estimated using Bayesian calibration and Monte Carlo simulations (see for example Van Oijen et al. 2011; Rahn et al. 2011).

Process-based models are available for the majority of biological GHG sources and sinks but tend to be limited to one type of source or sink. For instance, CENTURY, DAYCENT, and LandscapeDNDC (Giltrap et al. 2010; Haas et al. 2013) were developed to simulate biomass production and soil processes, including simulation of soil GHG fluxes and soil C/N stock changes. Process-based models are also available to simulate CH_4 emissions from livestock but have so far mainly been applied in the United States and in Europe (Thornton and Herrero 2010; Rotz et al. 2012; Duretz et al. 2011). These models are reasonable when evaluating the soil carbon sequestration potential at large scales or emissions of N_2O from monoculture fields (Babu et al. 2006), or changes in herd management (Pathak et al. 2005; Bryan et al. 2013; Perlman et al. 2013) but perhaps less so when trying to characterize the GHG impacts of smallholder systems at the whole farm level or for landscape-scale accounting.

Smallholder farming systems comprise multiple types of farming activities, often combining trees, animals, and crops in interconnected systems. Human management alters nutrient flows, potentially mitigating or exacerbating emissions from parts of the system; applying sectoral process-based models to whole farms therefore may oversimplify the complex interactions taking place (Tittonell et al. 2009). As of yet, few modeling approaches have been adapted for farm-level modeling of GHG impacts in mixed crop–livestock systems (Schils et al. 2007; Groot et al. 2012; Del Prado et al. 2013) and to our knowledge none have been applied to smallholder conditions of tropical developing countries.

To facilitate the widespread use of process-based models, as a first step the models need to be tested for most locations dominated by smallholder farming, which requires the availability of respective test datasets. Data on site-specific factors such as soil properties, cropping sequences, and fertilizer use are required; information which is often unavailable in many developing countries. In terms of enteric fermentation, the challenge is both a lack of information on animal numbers, species, and breeds, feeding regimes, as well as the quality of feeds and forages even though the models are based on the presumption that the chemical reactions that occur in the rumen are fairly standard and tend to go to completion. However, emission factors and rates currently available which have been obtained so far, don't consider that livestock production in developing countries often involves periods of severe undernutrition with feed qualities being far lower than tested in experiments in OECD countries. It is obvious that there is a great need to generate data that can be used for model parameterization and evaluation for smallholder conditions. Until now, only limited information has been available to independently assess the validity of the emission models for developing country conditions, casting doubt on the reliability of results generated from process-based models.

Conclusion

The complexity and scale that is characteristic of smallholder farming and the general lack of data presents significant challenges for scaling GHG emissions with much certainty. Significant efforts and investments are needed to improve systems representation so that the data collected are used to improve either empirical or process-based models. Moreover, conducting detailed monitoring campaigns can address the challenge of complexity and heterogeneity, and provide data that can be used to scale up representative systems with greater confidence.

Besides concerns over accuracy, technical demands in terms of data availability and model testing all limit the utility of using process-based models as a scaling method for GHG fluxes in agricultural systems of tropical developing countries at this time. However, given the costs of monitoring programs, it becomes an imperative to establish programs that can adapt and improve process-based models for quantification as they provide a means to test hypotheses of mitigation options and GHG accounting. This will require a number of investments in monitoring of smallholder practices of field and livestock management, scientific capacity building, and GHG measurements to evaluate the models for smallholder conditions. We estimate that a 10-year program of targeted and iterative measurements and modeling—those for key sources and sinks spanning heterogeneous conditions—is needed before use of process-based models becomes a viable solution for widespread GHG quantification in smallholder systems at either farm or landscape scales. In the meantime, models can be parameterized and tested well for farm and landscape situations, albeit time and resource intensive, but the limitations need to be recognized by those using the models and more importantly those using the model outputs.

References

Babu YJ, Li C, Frolking S, Nayak DR, Adhya TK (2006) Field validation of DNDC model for methane and nitrous oxide emissions from rice-based production systems of India. Nutr Cycl Agroecosyst 74:157–174

Blöschl G, Sivapalan M (1995) Scale issues in hydrological modelling: a review. Hydrol Process 9(3–4):251–290

Butterbach-Bahl K, Baggs EM, Dannenmann M, Kiese R, Zechmeister- Boltenstern S (2013) Nitrous oxide emissions from soils: how well do we understand the processes and their controls. Philos Trans R Soc B 368:20130122

Bryan E, Ringler C, Okoba B, Koo J, Herrero M, Silvestri S (2013) Can agriculture support climate change adaptation, greenhouse gas mitigation and rural livelihoods? Insights from Kenya. Clim Change 118:151–165

Colomb V, Bockel L (2013) Selection of appropriate calculators for landscape-scale greenhouse gas assessment for agriculture and forestry. Environ Res Lett 8:015029

De Gryze S, Wolf A, Kaffka SR, Mitchell J, Rolston DE, Temple SR, Lee J, Six J (2010) Simulating greenhouse gas budgets of four California cropping systems under conventional and alternative management. Ecol Appl 20:1805-1819. Del Grosso SL, Wirth T, Ogle SM, Parton WJ (2008) Estimating agricultural nitrous oxide emissions. EOS, Trans Am Geophys Union 89(51):529–540

Del Prado A, Crosson P, Olesen JE, Rotz C (2013) Whole-farm models to quantify greenhouse gas emissions and their potential use for linking climate change mitigation and adaptation in temperate grassland ruminant-based farming systems. Animal 7(Suppl 2):373–385

Duretz S, Drouet JL, Durand P, Hutchings NJ, Theobald MR, Salmon-Monviola J, Dragosits U, Maury O, Sutton MA, Cellier P (2011) Nitroscape: a model to integrate nitrogen transfers and transformations in rural landscapes. Environ Pollut 159:3162–3170

Ewert F, van Ittersum MK, Heckelei T, Therond O, Bezlepkina I, Andersen E (2011) Scale changes and model linking methods for integrated assessment of agri-environmental systems. Agr Ecosyst Environ 142:6–17

Ewert F (2004) Modelling changes in global regionalized food production systems under changing climate and consequences for food security and environment—development of an approach for improved crop modelling within IMAGE. Plant Production Systems Group, Department of Plant Sciences, Wageningen University & Netherlands Environmental Assessment Agency (MNP), National Institute for Public Health and Environment (RIVM), Wageningen, Netherlands

Gerber PJ, Steinfeld H, Henderson B, Mottet A, Opio C, Dijkman J, Falcucci A, Tempio G (2013) Tackling climate change through livestock—a global assessment of emissions and mitigation opportunities. Food and Agriculture Organization of the United Nations (FAO), Rome

Giltrap DL, Li C, Saggar S (2010) DNDC: a process-based model of greenhouse gas fluxes from agricultural soils. Agr Ecosyst Environ 136:292–300

Google Maps (2015) maps.google.com. Accessed 15 June 2015. Groffman PM, Butterbach-Bahl K, Fulweiler RW, Gold AJ, Morse JL, Stander EK, Tague CL, Tonitto C, Vidon P (2009) Challenges to incorporating spatially and temporally explicit phenomena (hotspots and hot moments) in denitrification models. Biogeochemistry 93:49–77

Groot JCJ, Oomen GJM, Rossing WAH (2012) Multi-objective optimization and design of farming systems. Agr Syst 110:63–77

Haas E, Klatt S, Fröhlich A, Kraft P, Werner C, Kiese R, Grote R, Breuer L, Butterbach-Bahl K (2013) Landscape DNDC: a process model for simulation of biosphere-atmosphere-hydrosphere exchange processes at site and landscape scale. Landscape Ecol 28:615–636

Hickman JE, Scholes RJ, Rosenstock TS, Garcia-Pando CP, Nyamangara J (2014) Assessing non-CO2 climate-forcing emissions and mitigation in sub-Saharan Africa. Curr Opin Environ Sustain 9–10:65–72

Hickman JE, Tully KL, Groffman PM, Diru W, Palm CA (2015) A potential tipping point in tropical agriculure: Avoiding rapid increases in nitrous oxide fluxes from agricultural intensification in Kenya. J Geo Res Bio. 120:1-10. Holling CS (1992) Cross-scale morphology, geometry, and dynamics of ecosystems. Ecol Monogr 62(4):447–502

IPCC (2006) 2006 IPCC guidelines for national greenhouse gas inventories. In: Eggleston S, Buendia L, Miwa K, Ngara T, Tanabe K (eds) Agriculture, forestry and other land use. IGES, Kanagawa

Li C, Frolking S, Xiao X, Moore III B, Boles S, Qiu J, Huang Y, Salas W, Sass R (2005) Modeling impacts of farming management alternatives on CO2, CH4, and N2O emissions: A case study for water management of rice agriculture in China. Glob Bio Cycles 19:1-10 Ogle SM, Buendia L, Butterbach-Bahl K et al (2013) Advancing national greenhouse gas inventories for agriculture in developing countries: improving activity data, emission factors and software technology. Environ Res Lett 8:015030

Owen JJ, Silver WL (2015) Greenhouse gas emissions from dairy manure management—a review of field based studies. Glob Chang Bio 21:550–565

Pathak H, Li C, Wassmann R (2005) Greenhouse gas emissions from Indian rice fields: calibration and upscaling using the DNDC model. Biogeosciences 2:113–123

Perlman J, Hijmans RJ, Horwath WR (2013) Modelling agricultural nitrous oxide emissions for large regions. Environ Model Software 48:183–192

Predotova M, Schlecht E, Buerkert A (2010) Nitrogen and carbon losses from dung storage in urban gardens of Niamey, Niger. Nutr Cycl Agroecosys 87:103–114

Rahn KH, Butterbach-Bahl K, Werner C (2011) Selection of likelihood parameters for complex models determines the effectiveness of Bayesian calibration. Ecol Inform 6:333–340

Rotz CA, Corson MS, Chianese DS, Montes F, Hafner SD, Jarvis R, Coiner CU (2012) Integrated farm system model: reference manual. USDA Agricultural Research Service, University Park. http://www.ars.usda.gov/Main/docs.htm?docid=21345. Accessed 8 Oct 2015

Schils RLM, Olesen JE, del Prado A, Soussana JF (2007) A review of farm level modelling approaches for mitigating greenhouse gas emissions from ruminant livestock systems. Livest Sci 112:240–251

Seebauer M (2014) Whole farm quantification of GHG emissions within smallholder farms in developing countries. Environ Res Lett 9:035006

Sieber S, Amjath-Babu TS, Jansson T, Müller K, Tscherning K, Graef F, Pohle D, Helming K, Rudloff B, Saravia-Matus BS, Gomez y Paloma S (2013) Sustainability impact assessment using integrated meta-modeling: simulating the reduction of direct support under the EU common agricultural policy. Land Use Policy 33:235–245

Shcherbak I, Millar N, Robertson GP (2014) Global metaanalysis of the nonlinear response of soil nitrous oxide emissions to fertilizer nitrogen. Proc Natl Acad Sci 111:9199–9204

Spencer S, Ogle SM, Breidt FJ, Goebel JJ, Paustian K (2011) Designing a national soil carbon monitoring network to support climate change policy: a case example for US agricultural lands. Greenhouse Gas Meas Manage 1:167–178

Stehfest E, Bouwman L (2006) N2O and NO emissions from agricultural fields and soils under natural vegetation: summarizing available measurement data and modeling of global annual emissions. Nutr Cycl Agroecosyst 74(3):207–228

Thornton PK, Herrero M (2010) Potential for reduced methane and carbon dioxide emissions from livestock and pasture management in the tropics. Proc Natl Acad Sci U S A 107: 19667–19672

Tittonell P, van Wijk MT, Herrero M, Rufino MC, de Ridder N, Giller KE (2009) Beyond resource constraints—exploring the biophysical feasibility of options for the intensification of smallholder crop-livestock systems in Vihiga District, Kenya. Agr Syst 101:1–19

Tubiello FN, Salvatore M, Rossi S, Ferrara A, Fitton N, Smith P (2013) The FAOSTAT database of greenhouse gas emissions from agriculture. Environ Res Lett 8:015009

Valentini R, Arneth A, Bombelli A et al (2014) A full greenhouse gases budget of africa: synthesis, uncertainties, and vulnerabilities. Biogeosciences 11:381–407

Van Delden H, van Vliet J, Rutledge DT, Kirkby ML (2011) Comparison of scale and scaling issues in integrated land-use models for policy support. Agr Ecosyst Environ 142(1–2):18–28

Van Groenigen JW, Velthof GL, Oeneme O, Van Groenigen KJ, Van Kessel C (2010) Towards an agronomic assessment of N2O emissions: a case study for arable crops. Eur J Soil Sci 61:903–913

Van Oijen M, Cameron DR, Butterbach-Bahl K, Farahbakhshazad N, Jansson PE, Kiese R, Rahn KH, Werner C, Yeluripati JB (2011) A Bayesian framework for model calibration, comparison and analysis: application to four models for the biogeochemistry of a Norway spruce forest. Agric Forest Met 151(12):1609–1621

Vermeulen SJ, Campbell BM, Ingram JSI (2012) Climate change and food systems. Annu Rev Environ Res 37:195–222

Volk M, Ewert F (2011) Scaling methods in integrated assessment of agricultural systems—state of the art and future directions. Agr Ecosyst Environ 142:1–5

Chapter 10
Methods for Environment: Productivity Trade-Off Analysis in Agricultural Systems

Mark T. van Wijk, Charlotte J. Klapwijk, Todd S. Rosenstock, Piet J.A. van Asten, Philip K. Thornton, and Ken E. Giller

Abstract Trade-off analysis has become an increasingly important approach for evaluating system level outcomes of agricultural production and for prioritising and targeting management interventions in multi-functional agricultural landscapes. We review the strengths and weakness of different techniques available for performing trade-off analysis. These techniques, including mathematical programming and participatory approaches, have developed substantially in recent years aided by mathematical advancement, increased computing power, and emerging insights into systems behaviour. The strengths and weaknesses of the different approaches are identified and discussed, and we make suggestions for a tiered approach for situations with different data availability. This chapter is a modified and extended version of Klapwijk et al. (2014).

M.T. van Wijk (✉)
International Livestock Research Institute, Old Naivasha Rd., P.O. Box 30709, Nairobi, Kenya
e-mail: m.vanwijk@cgiar.org

C.J. Klapwijk
Plant Production Systems Group, Wageningen University and Research Centre, Wageningen, Netherlands

International Institute of Tropical Agriculture (IITA), Kampala, Uganda
e-mail: l.klapwijk@cgiar.org

T.S. Rosenstock
World Agroforestry Centre (ICRAF), Nairobi, Kenya

P.J.A. van Asten
International Institute of Tropical Agriculture (IITA), Kampala, Uganda

P.K. Thornton
CGIAR Research Program on Climate Change, Agriculture and Food Security (CCAFS), Nairobi, Kenya

K.E. Giller
Plant Production Systems Group, Wageningen University, Wageningen, Netherlands

T.S. Rosenstock et al. (eds.), *Methods for Measuring Greenhouse Gas Balances and Evaluating Mitigation Options in Smallholder Agriculture*,
DOI 10.1007/978-3-319-29794-1_10

10.1 Introduction

Trade-offs, by which we mean exchanges that occur as compromises, are ubiquitous when land is managed with multiple goals in mind. Trade-offs may become particularly acute when resources are constrained and when the goals of different stakeholders conflict (Giller et al. 2008). In agriculture, trade-offs between output indicators may arise at all hierarchical levels, from the crop (such as grain versus crop residue production), the animal (milk versus meat production), the field (grain production versus nitrate leaching and water quality), the farm (production of one crop versus another), to the landscape and above (agricultural production versus land for nature). An individual farmer may face trade-offs between maximizing production in the short term and ensuring sustainable production in the long term. Within landscapes, trade-offs may arise between different individuals for competing uses of land. Thus trade-offs exist both within agricultural systems, between agricultural and broader environmental or sociocultural objectives, across time and spatial scales, and between actors. Understanding the system dynamics that produce and change the nature of the trade-offs is central to achieving a sustainable and food-secure future.

In this chapter we focus on how the complex relationships between the management of farming systems and its consequences for production and the environment—here represented by greenhouse gas (GHG) emissions—can be analyzed and how trade-offs and possible synergies between output indicators can be quantified. For example, an important hypothesis is that by increasing soil carbon sequestration in agricultural systems, farmers can generate a significant share of the total emission reductions required in the next few decades to avoid catastrophic levels of climate change. At the same time, increasing soil carbon sequestration also increases soil organic matter, which is fundamental to improving the productivity and resilience of cropping and livestock production systems, and thereby a potential win–win situation is identified. However, it is debatable whether these win–win situations exist in reality. An important constraint for this hypothesis is the lack of organic matter like crop residues on many smallholder mixed crop–livestock systems, to serve both as feed for livestock and as an input into the soil in order to increase soil organic matter. This organic matter could be produced through the use of mineral fertiliser or intensification of livestock production, but both of these have negative consequences for GHG emissions, probably offsetting the gains made in soil organic matter storage. It therefore seems likely that to achieve maximum impact on smallholders' food production and food security, environmental indicators have to be compromised. However, good quantitative insight into these compromises is still lacking.

Trade-off analysis has emerged as one approach to assessing farming system dynamics from a multidimensional perspective. Although the concept of trade-offs and their opposite—synergies—lies at the heart of several recent agricultural research-for- development initiatives (Vermeulen et al. 2011; DeFries and Rosenzweig 2010), methods to analyze trade-offs within agroecosystems and the wider landscape are nascent (Foley et al. 2011). We review the state of the art for trade-off analyses, highlighting important innovations and constraints, and discuss the strengths and weaknesses of the different approaches used in the recent literature.

10.2 The Nature of Trade-Off Analysis

Trade-offs are quantified through the analysis of system-level inputs and outputs such as crop production, household labour use, or environmental impacts such as greenhouse gas emissions. The outcomes that different actors may want to achieve, in and beyond the landscape, need to be defined at different time and spatial scales. Understanding these desired outcomes, or different stakeholders' objectives, is a necessary first step in trade-off analysis.

We illustrate the key concepts and processes of trade-off analysis with a simple example that has only two objectives: farm-scale production and an environmental impact on greenhouse gas emissions. Once the objectives have been defined, the next step is to identify meaningful indicators that describe these objectives. The indicators form the basis for characterizing the relationships between objectives (Fig. 10.1). The shape of the trade-off curve gives important information on the severity of the trade-off of interest. Is it simply a straight line, like the central curve (Fig. 10.1a)? Is the curve convex (i.e. the lower curve), which means strong trade-offs exist between the indicators); or concave (i.e. the upper curve), which means the indicators are independent of each other and the trade-offs between the indicators are quite 'soft'? The shape of the trade-off curve represents different functional relationships and can be assessed by evaluating farm management options; in our example, each point could represent a method and level of mineral fertiliser application (Fig. 10.1b). The position of each option in the trade-off space describes its outcomes in terms of the two indicators, productivity and environmental impact. Based on this information, a 'best' trade-off curve can be drawn (Fig. 10.1c). In trade-off analyses the researcher will be interested in which system management interventions result in which type of outcome of the different objectives (Fig. 10.1d).

Once the best (observed or inferred) trade-off curve has been identified, various system management interventions can be studied to assess the extent to which they contribute to the desired objectives (Fig. 10.1d). This analysis determines whether so-called 'win–win' solutions are possible, where the performance of the system can be improved with regard to both objectives. Alternatively, does improvement in one objective automatically lead to a decrease in system performance for another objective (Fig. 10.1e)? Possible threshold values can be identified once the shape of the trade-off curve is known. For example, do productivity thresholds exist, above which the environmental impact increases rapidly? In some situations, it may be possible to alter the nature of the trade-off between production and environmental impact through the exploration of new management interventions (Fig. 10.1f), thereby redefining the 'best' trade-off curve.

10.3 Research Approaches and Tools

Trade-offs are typically much more complex with more dimensions and objectives than indicated by the simple two-dimensional examples presented in the previous section. A wide variety of tools and approaches have been developed to account for

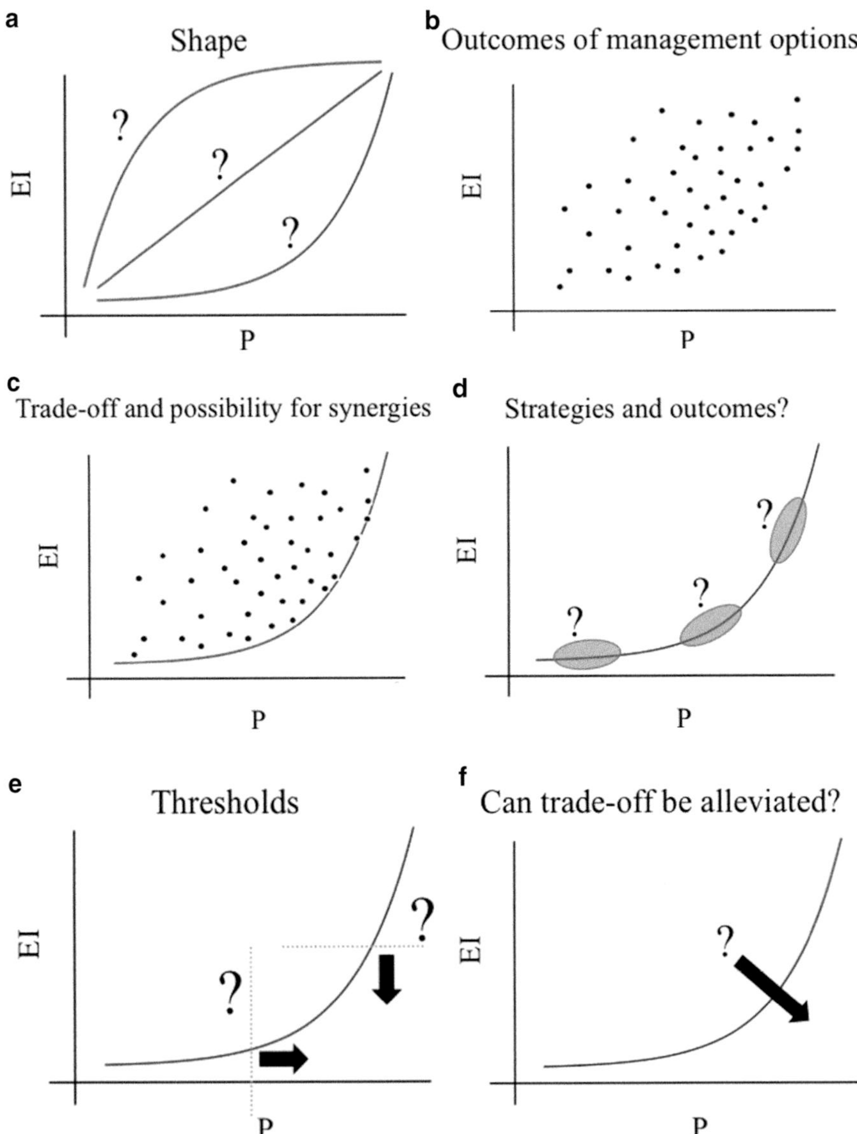

Fig. 10.1 Key concepts of trade-offs and their analysis of a simple two-objective example (for explanation see text) EI = environmental impact, P = production. (**a**) Shape, (**b**) outcomes of management options, (**c**) trade-off and possibility for synergies, (**d**) strategies (interventions) and outcomes, (**e**) thresholds, (**f**) can trade-offs be alleviated

diverse situations. The most suitable approach depends on the nature and scale of the problem to be addressed, the trade-offs involved, and the indicators available. We assess five widely applied approaches: (1) participatory methods; (2) empirical analyses; (3) econometric tools; (4) optimization models, and (5) simulation models. These five approaches overlap often and can help generate complementary knowledge. Consequently, trade-off analyses will often utilize several methods simultaneously or iteratively.

The concept of *participatory research* originally highlighted the need for the active involvement of those who are the subject of research, or for whom the research may lead to outcome changes. In recent times, the notion has expanded to acknowledge that change in researchers' assumptions and perceptions may be required to achieve desired outcomes that are attractive to farmers (Crane 2010). Participatory approaches, such as fuzzy cognitive mapping (Murungweni et al. 2011), resource flow mapping, games and role-playing, are powerful ways to identify actor-relevant objectives and indicators, although the scope of farmer knowledge and perceptions within scientific research can be constraining in some situations, particularly in times of rapid change (Van Asten et al. 2009). There are many examples of participatory approaches (Gonsalves 2013) that could be or are used to assess trade-offs. Participatory approaches usually generate qualitative data and so, although they may not be well suited for quantifying trade-offs, they provide critically important information to support quantitative tools, for example through the development of participatory scenarios (DeFries and Rosenzweig 2010; Claessens et al. 2012). However, despite the participatory nature of these approaches, the assessment of trade-offs often remains researcher-driven.

Quantitative assessment of trade-offs requires *empirical* or experimental approaches to generate data on the behavior of the system under different conditions. Trade-off curves can be drawn on the basis of experimental measurements of indicators, such as the removal of plant biomass for fodder and the resulting soil cover, which is a good proxy for control of soil erosion (Naudin et al. 2012). Statistical techniques such as data envelope analysis (Fraser and Cordina 1999) or boundary line analysis (Fermont et al. 2009) can be used to quantify best possible trade-offs between indicators in empirical datasets (e.g. Fig. 10.1c). Related to these empirical approaches are *econometric tools*: these use large datasets as the basis of statistical coefficients that define the input–output relationships of system level outcomes (e.g. Antle and Capalbo 2001). Developments combine biophysical and socioeconomic aspects of the system, and use farm-level responses to quantify consequences at a regional level (Antle and Stoorvogel 2006). Empirical and econometric approaches are powerful in the sense that outcomes of various system choices can be explored using the existing variability in system configuration and performance. However, the inference space of the analysis is constrained to the dataset collected and is therefore not suitable for predicting outcomes outside the ranges of the original data.

Empirical approaches cannot be used to assess indicators that are difficult to measure directly; therefore, they are often combined with *simulation models* to obtain an overview of overall system performance. Simulation models allow the dynamic nature of trade-offs to be explored, where outcomes can differ in the short or long term (Zingore et al. 2009). System performance, expressed quantitatively in terms of outcomes represented by different indicators, can be used as an input for *optimization* approaches such as mathematical programming (MP). MP finds the best possible trade-off through multicriteria analysis and can assess whether this trade-off curve can be alleviated through new interventions. MP has a long history (e.g. Hazell and Norton 1986) and is among the most extensively used trade-off application in land use studies (e.g. Janssen and Van Ittersum 2007). This is despite its inherent limitation, that land users do not always behave according to economic rationality and optimise their behaviour. Techniques have been developed recently to solve non-linear MP problems and integrate across levels, linking farms and regions through markets and environmental feedbacks (e.g. Laborte et al. 2007; Roetter et al. 2007; Louhichi et al. 2010).

Inverse modelling techniques use non-linear *simulation models* directly to perform multiobjective optimization without the intermediate step of MP. Furthermore, with the identification of the appropriate model outputs, system behaviour can be assessed across different temporal and spatial scales and feedbacks taken into account, which is often a weak part of MP models. The complexity of agroecosystems and the large number of potential indicators can hamper efficient applications of this computationally intensive method. But advances in computer power have resulted in several applications in farming systems research, going from farm to landscape (Groot et al. 2007, 2012; Tittonell et al. 2007).

The various approaches to trade-off analysis each have key strengths and weaknesses and combining approaches may provide enhanced opportunities for a realistic, relevant, and integrated assessment of systems (Table 10.1). For example, in many cases, participatory approaches are needed to define meaningful objectives and indicators, but are not suitable to reliably quantify the trade-offs associated with possible interventions. Empirical and econometric approaches can be used to quantify the current state of the overall agricultural system. In many cases, however, simulation models are needed to quantify indicators that are difficult to measure (for example, effects of management on longer term productivity) and to explore options beyond the existing system configurations and boundaries (Table 10.1). Optimization can be used to assess the potential for synergies and alleviation of trade-offs, but has limited applicability when sociocultural traditions and rules play a key role (Thornton et al. 2006).

It is clear that for trade-off analyses combinations of techniques are needed. Multicriteria analysis is an example of such an integrated approach, in which participatory and optimization methods are combined: the weighting of the individual criteria in goal programming models is done together with the stakeholders, and by changing these weights with the stakeholders a trade-off analysis is performed (e.g. Romero and Rehman 2003).

10.4 A Tiered Approach

The discussion above demonstrates that for fully integrated trade-off analyses different approaches should be combined. However, in many cases data availability will not allow such elaborate analyses. The techniques discussed in the previous

Table 10.1 Strengths and weaknesses of the different approaches for analysing trade-offs in agricultural systems

	Research approach									
	Empirical		Econometric		Simulation		Optimization		Participatory	
Aspect	Act	Pot	Act	Pot	Act	Pot	Act	Pot	Act	Pot
Integration of interdisciplinary content	–	+	+	+	–	+	–	–	–	+
Assessment across different time horizons	–	–	+	+	+	+	+	+	–	+
Assessment across spatial scales and integration levels	–	+	–	+	+/–	+/–	+/–	+	–	+
Takes into account qualitative information	–	+	–	–	–	–	–	–	+	+
Appropriate representation of uncertainty	–	+	–	+	–	+	–	+	–	+
Identification of possibilities to alleviate the observed trade-offs	–	–	–	–	+	+	+	+	–	–
Ability to deal with real-life system complexity	+	+	–	+	–	–	–	–	+	+
Applicability to real-life decision-making	+	+	+	+	–	–	+/–	+/–	+	+

Act actual or current use in the scientific literature, *Pot* potential usefulness of technique

section not only have different strengths and weaknesses, but also different data demands. Typically, empirical and econometric approaches are highly data-demanding, and therefore time-consuming and expensive, whereas participatory approaches can provide essential information about system functioning after only a few well-designed discussion panels and targeted questionnaires. Simulation and optimization models can be, in terms of data demand, anywhere between these extremes. Their data demand is highly determined by model setup and complexity.

An example of a tiered approach in which researchers move from quick initial data analyses to more complex, data demanding, modelling exercises is the four-step approach used by Van Noordwijk and his team at ICRAF (Meine van Noordwijk, personal communication; see also Tata et al. 2014 for the first three steps; Villamor et al. 2014 for an agent-based modelling approach).

- Step one is the collection of system characterization data and the analysis of these data to explore whether trade-offs can be identified, for example between an environmental indicator like soil carbon and the net present value of the land.
- The second step is to look at these variables from a dynamic perspective and identify opportunities for interventions by analysing the opportunity costs of different management options. This step already requires much more detailed data than step 1, and in the example above, could be used to identify the price of emission reduction potentials.
- In the third step, the consequences of the identified intervention options for the different land users and the environment can be explored by using dynamic land-use models.
- Finally in the fourth step, agent-based models and participatory modelling exercises are used to analyse the opinions of, and interactions between, different actors in the landscape. This provides an integrated analysis of both the environmental and socioeconomic factors and actors within the landscape.

This four-step approach demonstrates the way in which the strengths of different methods of trade-off analysis can be combined, and how such an analysis can move stepwise towards more complex and data-demanding exercises.

All in all it is not straightforward to give concrete advice that relates the purpose of analysis to the technique and approach to be used. Researchers make personal choices about complexity and analytical approach as part of the 'art' of modelling and trade-off analyses. This is sometimes difficult to reconcile with the 'objectivity' that we pursue in scientific research. However, some general indications can be given.

If the objective of the analysis is to assess the overall potential for system improvement and the room for manoeuvre to increase efficiencies and profitability without negative effects on environmental indicators, then optimization approaches are the most logical choice. If the purpose is to analyse the short- and long-term consequences of certain interventions and the trade-offs between different objectives over different time scales, then simulation modelling is an obvious candidate. This may be combined with some sort of multiobjective, non-linear optimization or inverse modelling approach.

Both optimization and simulation are typically used for scientifically oriented studies. In order to have real-life impact, that takes into account the complexities of agricultural systems and the large diversity of drivers and options in agricultural land use, especially in developing countries, a variety of quantitative and qualitative approaches are likely to be needed (e.g. Murungweni et al. 2011). The setup of these tools, the identification of indicators, and the presentation of results need to be determined using participatory approaches where key stakeholders are involved and drive decisions from the beginning of the project. This might lead to the study having less value in terms of scientific novelty, but will increase its practical relevance on the ground. With the topic of this chapter in mind, it is ironic that in many cases there might be a trade-off between the scientific and societal impact that can be achieved by a research project that has its own constraints in terms of time and money.

Acknowledgements This study is an outcome of a workshop entitled 'Analysis of Trade-offs in Agricultural Systems' organised at Wageningen University, February 2013. We thank all participants for their discussions, which contributed strongly to the content of this chapter. The workshop and subsequent work were funded by the CGIAR Research Program on Climate Change, Agriculture and Food Security (CCAFS), Theme 4.2: *Integration for Decision-Making—Data and Tools for Analysis and Planning*. This chapter is a modified and extended version of Klapwijk et al. (2014).

References

Antle JM, Capalbo SM (2001) Econometric-process models for integrated assessment of agricultural production systems. Am J Agric Econ 83:389–401

Antle JM, Stoorvogel JJ (2006) Incorporating systems dynamics and spatial heterogeneity in integrated assessment of agricultural production systems. Environ Dev Econ 11:39–58

Claessens L, Antle JM, Stoorvogel JJ, Valdivia RO, Thornton PK, Herrero M (2012) A method for evaluating climate change adaptation strategies for small-scale farmers using survey, experimental and modelled data. Agr Syst 111:85–95

Crane T (2010) Of models and meanings: cultural resilience in social-ecological systems. Ecol Soc 15:19

DeFries R, Rosenzweig C (2010) Toward a whole-landscape approach for sustainable land use in the tropics. Proc Natl Acad Sci U S A 107:19627–19632

Fermont AM, Van Asten PJA, Tittonell PA, Van Wijk MT, Giller KE (2009) Closing the cassava yield gap: an analysis from smallholder farmers in East Africa. Field Crop Res 112:24–36

Foley JA, Ramankutty N, Brauman KA, Cassidy ES, Gerber JS, Johnston M, Mueller ND, O'Connell C, Ray DK, West PC, Baizer C, Bennett EM, Carpenter SR, Hill J, Monfreda C, Polasky S, Rockström J, Sheehan J, Siebert S, Tilman D, Zaks PM (2011) Solutions for a cultivated planet. Nature 478:337–442

Fraser I, Cordina D (1999) An application of data envelopment analysis to irrigated dairy farms in Northern Victoria, Australia. Agr Syst 59:267–282

Giller KE, Leeuwis C, Andersson JA, Andriesse W, Brouwer A, Frost P, Hebinck P, Heitkönig I, Van Ittersum MK, Koning N, Ruben R, Slingerland M, Udo H, Veldkamp T, van de Vijver C, van Wink MT, Windmeijer P (2008) Competing claims on natural resources: what role for science? Ecol Soc 13:34

Gonsalves JF (2013) A new relevance and better prospects for wider uptake of social learning within the CGIAR. CCAFS working paper no. 37. CGIAR Research Program on Climate Change, Agriculture and Food Security (CCAFS), Copenhagen, Denmark. http://www.ccafs.cgiar.org/. Accessed 13 Mar 2015

Groot JCJ, Oomen GJM, Rossing WAH (2012) Multi-objective optimization and design of farming systems. Agr Syst 110:63–77

Groot JCJ, Rossing WAH, Jellema A, Stobbelaar DJ, Renting H, Van Ittersum MK (2007) Exploring multi-scale trade-offs between nature conservation, agricultural profits and landscape quality—a methodology to support discussions on land-use perspectives. Agr Ecosyst Environ 120:58–69

Hazell PBR, Norton RD (eds) (1986) Mathematical programming for economic analysis in agriculture. Macmillan, New York, p 400

Janssen S, Van Ittersum MK (2007) Assessing farm innovations and responses to policies: a review of bio-economic farm models. Agr Syst 2007(94):622–636

Klapwijk CJ, van Wijk MT, Rosenstock TS, van Astern PJA, Thornton PK, Giller KE (2014) Analysis of trade-offs in agricultural systems: current status and way forward. Curr Opin Environ Sustain 6:110–115

Laborte AG, Van Ittersum MK, Van den Berg MM (2007) Multi-scale analysis of agricultural development: a modelling approach for Ilocos Norte, Philippines. Agr Syst 94:862–873

Louhichi K, Flichman G, Boisson JM (2010) Bio-economic modelling of soil erosion externalities and policy options: a Tunisian case study. J Bioecon 12:145–167

Murungweni C, Van Wijk MT, Andersson JA, Smaling EC, Giller KE (2011) Application of fuzzy cognitive mapping in livelihood vulnerability analysis. Ecol Soc 16(4):8

Naudin K, Scopel E, Andriamandroso ALH, Rakotosolofo M, Andriamarosoa Ratsimbazafy NRS, Rakotozandriny JN, Salgado P, Giller KE (2012) Trade-offs between biomass use and soil cover. The case of rice-based cropping systems in the Lake Alaotra region of Madagascar. Exp Agric 48:194–209

Roetter RP, Van den Berg M, Laborte AG, Hengsdijk H, Wolf J, Van Ittersum M, Van Keulen H, Agustin EO, Son TT, Lai NX, Guanghuo W (2007) Combining farm and regional level modelling for integrated resource management in East and South-east Asia. Environ Model Software 22:149–157

Romero C, Rehman T (eds) (2003) Multiple criteria analysis for agricultural decisions. Elsevier, Amsterdam

Tata HL, van Noordwijk M, Ruysschaert D, Mulia R, Rahayu S, Mulyoutami E, Widayati A, Ekadinata A, Zen R, Darsoyo A, Oktaviani R, Dewi S (2014) Will funding to reduce emissions from deforestation and (forest) degradation (REDD+) stop conversion of peat swamps to oil palm in orangutan habitat in Tripa in Aceh, Indonesia? Mitig Adapt Strat Glob Chang 19:693–713

Thornton PK, Burnsilver SB, Boone RB, Galvin KA (2006) Modelling the impacts of group ranch subdivision on households in Kajiado, Kenya. Agr Syst 87:331–356

Tittonell PA, Van Wijk MT, Rufino MC, Vrugt JA, Giller KE (2007) Analysing trade-offs in resource and labour allocation by smallholder farmers using inverse modelling techniques: a case-study from Kakamega district, western Kenya. Agr Syst 95:76–95

Van Asten PJA, Kaaria S, Fermont AM, Delve RJ (2009) Challenges and lessons when using farmer knowledge in agricultural research and development projects in Africa. Exp Agric 45:1–14

Vermeulen S, Zougmoré R, Wollenberg E, Thornton PK, Nelson G, Kristjanson P, Kinyangi J, Jarvis A, Hansen J, Challinor AJ, Campbell B, Aggarwal P (2011) Climate change, agriculture and food security: a global partnership to link research and action for low-income agricultural producers and consumers. Curr Opin Environ Sustain 4:1–6

Villamor GB, Le QB, Djanibekov U, van Noordwijk M, Vlek PLG (2014) Biodiversity in rubber agroforests, carbon emissions, and rural livelihoods: an agent-based model of land-use dynamics in lowland Sumatra. Environ Model Software 61:151–165

Zingore S, González-Estrada E, Delve RJ, Herrero M, Dimes JP, Giller KE (2009) An integrated evaluation of strategies for enhancing productivity and profitability of resource-constrained smallholder farms in Zimbabwe. Agr Syst 101:57–68

Index

T.S. Rosenstock et al. (eds.), *Methods for Measuring Greenhouse Gas Balances and Evaluating Mitigation Options in Smallholder Agriculture*,
DOI 10.1007/978-3-319-29794-1